T0136847

Printed in the United States
by Baker & Taylor Publisher Services

Mechanisms and Machine Science

Volume 82

This book series establishes a well-defined forum for monographs, edited Books, and proceedings on mechanical engineering with particular emphasis on MMS (Mechanism and Machine Science). The final goal is the publication of research that shows the development of mechanical engineering and particularly MMS in all technical aspects, even in very recent assessments. Published works share an approach by which technical details and formulation are discussed, and discuss modern formalisms with the aim to circulate research and technical achievements for use in professional, research, academic, and teaching activities.

This technical approach is an essential characteristic of the series. By discussing technical details and formulations in terms of modern formalisms, the possibility is created not only to show technical developments but also to explain achievements for technical teaching and research activity today and for the future.

The book series is intended to collect technical views on developments of the broad field of MMS in a unique frame that can be seen in its totality as an Encyclopaedia of MMS but with the additional purpose of archiving and teaching MMS achievements. Therefore, the book series will be of use not only for researchers and teachers in Mechanical Engineering but also for professionals and students for their formation and future work.

The series is promoted under the auspices of International Federation for the Promotion of Mechanism and Machine Science (IFToMM).

Prospective authors and editors can contact Mr. Pierpaolo Riva (publishing editor, Springer) at: pierpaolo.riva@springer.com

Indexed by SCOPUS and Google Scholar.

More information about this series at http://www.springer.com/series/8779

Michał Ciszewski · Mariusz Giergiel ·
Tomasz Buratowski · Piotr Małka

Modeling and Control of a Tracked Mobile Robot for Pipeline Inspection

Springer

Michał Ciszewski
Department of Robotics and Mechatronics
AGH University of Science and Technology
Krakow, Poland

Mariusz Giergiel
Department of Robotics and Mechatronics
AGH University of Science and Technology
Krakow, Poland

Tomasz Buratowski
Department of Robotics and Mechatronics
AGH University of Science and Technology
Krakow, Poland

Piotr Małka
Department of Robotics and Mechatronics
AGH University of Science and Technology
Krakow, Poland

ISSN 2211-0984 ISSN 2211-0992 (electronic)
Mechanisms and Machine Science
ISBN 978-3-030-42717-7 ISBN 978-3-030-42715-3 (eBook)
https://doi.org/10.1007/978-3-030-42715-3

This Springer imprint is published by the registered company Springer Nature Switzerland AG
The registered company address is: Gewerbestrasse 11, 6330 Cham, Switzerland

Preface

In this book, design, mathematical modeling, control system development and experimental validation of a versatile pipe inspection mobile robot is presented. The design is based on an original, patented structure of mobile motion unit that features two track drives, integrated in a chassis, adjusted by six servomotors. This advantageous structure allows to use a single robotic unit that can be actively adapted for inspection and visual monitoring of different industrial facilities, including pipelines of various sizes and shapes, oriented horizontally and vertically, even or rough surfaces. Mathematical modeling of the robot includes kinematic and dynamic models of the robot motion, dedicated for inspection of even surfaces. The adaptive chassis of the robot, composed of two pedipulators with closed kinematic chains, requires complex modeling and control methods to realize its tasks. To transform the robot to diverse work environments, an original trajectory generation algorithm is developed that utilizes 3D model of the robot, analytical and numerical forward and inverse kinematics with addition of custom rules. The mathematical models are verified by dedicated simulations of pedipulators transformation and adaptation to different pipe sizes, prepared in the MATLAB software. To predict real operation of a prototype, extensive verification of the robot model is successfully conducted by custom co-simulations that involve usage of the MATLAB/Simulink software for mathematical model processing, controller design, hardware interaction and the V-REP application for simulation of the 3D robot model and execution of control commands. Control system of the robot is designed as a multidisciplinary solution, based on methods applied for mobile robots and arm-type robots. Development of a custom electronic board is discussed with focus on functionality and application of the control system. Software development covers low-level applications for the robot on-board controller and high-level operator program that generates commands, allows adaptation of the robot chassis and visualizes inspection data. Development of a prototype of the robot is presented, including integration of mechanical, electrical and electronic components with sensors and the control system. Laboratory experiments are conducted that verify the modeling and control system design in different work scenarios and operating conditions. These tests show that the thesis of this book is fulfilled. An enhanced version of the prototype is presented with augmented functionality in the aspect of mechanical, communication, power supply and control systems that is ready for industrial use.

Krakow, Poland

Michał Ciszewski
Mariusz Giergiel
Tomasz Buratowski
Piotr Małka

Contents

Abbreviations

2D	Two-Dimensional
3D	Three-Dimensional
AC	Alternating Current
API	Application Programming Interface
CAD	Computer Aided Design
CAM	Computer Aided Manufacturing
CCTV	Closed-Circuit TeleVision
CNC	Computer Numerical Control
CSV	Comma Separated Value
DC	Direct Current
DH	Denavit-Hartenberg notation
DOF	Degrees Of Freedom
EDM	Electro-Discharge Machining
EKF	Extended Kalman Filter
FIR	Finite Impulse Response (filter)
GIS	Geographic Information System
GPS	Global Positioning System
IMU	Inertial Measurement Unit
IP address	Internet Protocol address
IP rating	Ingress Protection rating
IR camera	InfraRed camera
LED	Light Emitting Diode
LIDAR	LIght Detection And Ranging
MATLAB	MATrix LABoratory (software by MathWorks)
MFL	Magnetic Flux Leakage
NDE	Non-Destructive Evaluation
NDT	Non-Destructive Testing
PC	Personal Computer
PCB	Printed Circuit Board
PD controller	Proportional-Derivative controller

PID controller	Proportional-Integral-Derivative controller
PMMA	Poly-Methyl Methacrylate
PTZ	Pan-Tilt-Zoom
PVC	Poly-Vinyl Chloride
PWM	Pulse Width Modulation
SLAM	Simultaneous Localization And Mapping
TCP/IP	Transmission Control Protocol/Internet Protocol
TVL	TeleVision Lines
UAV	Unmanned Aerial Vehicle
UGV	Unmanned Ground Vehicle
USB	Universal Serial Bus
V-REP	Virtual Robot Experimental Platform (software by Coppelia Robotics)

Symbols

a_{f1}, a_{f2}	Lengths of pedipulator links 1, 2
$a_{r1}, a_{r2},\ a_{r3}$	Lengths of pedipulator links 1, 2, 3
b	Width of the crawler track
B	Joint damping coefficients (the same for all joints)
C_{ij}	Coefficient matrix
dL	Track tread deformation
$\dot{\mathrm{e}}_{\mathrm{i}}$	Kinetic parameters in generalized coordinates
E	Kinetic energy
E_{K1}	Kinetic energy of the track drive sprocket 1
E_{K2}	Kinetic energy of the idler 2
E_{K3}	Kinetic energy of the idler 3
E_{M1}	Kinetic energy of the left track drive module
E_{M2}	Kinetic energy of the right track drive module
E_O	Kinetic energy of the track module housing
E_R	Kinetic energy of the robot frame
E_{TOT}	Total error of electric current sensors
F_D	Hydrostatic resistance force
F_E	External end-effector forces
F_w	Buoyant force
$f_x,\ f_y,\ f_z$	External generalized force components (forces)
G	Gravity force
G_E	Generalized external forces applied to move the end-effector to desired pose
G_j	Coefficients for Maggi's equations
H	Distance between the tracks
$i = 1,\ldots,s$	Number of independent parameters, expressed in generalized coordinates q_j, in the number equal to degrees of freedom of the system (Sects. 4.1 and 4.2)
i	Class of the kinematic pair (Sects. 4.3–4.5)
i	Index of joint in kinematic structure (Sects. 4.3–4.5)

i_{gear}	Internal meshing gears transmission ratio
I	Identity matrix
I_R	Moment of inertia of the robot frame
I_{SR}	Measurement range of a current sensor
I_X, I_Z	Reduced moments of inertia of the track drive module
I_{xi}	Moment of inertia with respect to i-th axis of rotation x
I_{xO}	Moment of inertia of the elements in rotational motion
I_{zi}	Moment of inertia of i-th wheel with respect to the axis z
I_{zO}	Moment of inertia of the housing with respect to the instantaneous center of rotation
$j = 1, .., n$	Number of generalized coordinates
J	Manipulator Jacobian matrix
J^+	Pseudo-inverse of the Jacobian J matrix
J_v	Jacobian elements representing linear velocities
J_ω	Jacobian elements representing angular velocities
k	Versor of z_0 axis
L	Length of the load-bearing segment of the crawler track
m	Mass of the track
m_{Ki}	Mass of i-th wheel
m_O	Mass of the track module housing
m_R	Mass of the robot frame
m_x, m_y, m_z	External generalized force components (moments)
M	Mask vector for inverse kinematics calculations
M_E	External end-effector moments
M_n	Torque on the drive sprocket
M_{n1}, M_{n2}	Torque on the drive sprockets of the tracks 1, 2
M_P	Moment of transverse resistance
M_R	Mask matrix for calculation of inverse kinematics with manipulator Jacobian
M_s	Track motor torque
n	Number of track treads in contact with the ground (Sects. 4.1 and 4.2)
n	Number of links in kinematic structure (Sects. 4.3–4.5)
p_i	Number of kinematic pairs of i-th class
$p_{n,0}(q)$	3×1 position vector
P_u	Tether cable pull force
p_x, p_y, p_z	Position with respect to x, y, z axis—respectively
P_n	Driving force
q	Vector of generalized joint variables
$\dot{q}$	Velocity of generalized joint coordinates (linear or angular)
$q\langle k\rangle$	Current estimate of inverse kinematics solution
q_f	Front manipulator drives angular positions
q_p	Pedipulator drives angular positions
q_r	Rear manipulator drives angular positions

Q	Generalized joint forces
r	Radius of a track drive sprocket
r_1	Radius of the sprocket 1
$r_{11} \ldots r_{33}$	Orientation of n-th coordinate system with respect to base coordinate system
r_C	Radius of circle of pipe cross-section
R_{AD}	Resolution of A/D transducer in the development board
R_D	Resolution of multimeter display
$R(t)$	3×3 time-varying rotation matrix
$R_{n,0}(q)$	3×3 rotation matrix
s	Number of degrees of freedom (Sects. 4.3 and 4.4)
s	Slip of a track (Sects. 4.1 and 4.2)
$s_1,\ s_2$	Slip of the tracks 1, 2
S	Sensitivity of current sensor
$S(\omega)$	Skew-symmetric matrix
$T_{E,0}$	Transformation matrix from end-effector to base coordinate system
$T_{n,0}$	Homogeneous transformation matrix from n-th to 0-th (base) coordinate system
U_M	Value of measured voltage
U_B	Input voltage range of development board
V	Spatial velocity
$v_{n,0}$	Linear velocity with respect to the base coordinate system
v_x, v_y, v_z	Linear velocities with respect to axes x, y, z
$V_A,\ V_B,\ V_C, V_E$	Velocities of characteristic points of the robot
w	Mobility of a mechanism
$W_t,\ W_{t1},\ W_{t2}$	Rolling friction forces
x	Distance of the point for which the slip is calculated from the point of crawler track contact with the ground
$x_0,\ y_0,\ z_0$	Axes of the base coordinate system
x_1, x_2, x_3	x axis coordinates of track contact points 1, 2, 3
x_C, y_C	Circle center point coordinates of pipe cross-section (Sect. 4.5)
$\dot{x}_C, \dot{y}_C,\ \dot{z}_C$	Velocity components of point C with respect to axes x, y, z (Sects. 4.1 and 4.2)
y_1, y_2, y_3	y axis coordinates of track contact points 1, 2, 3
z_i	Versor of axis Z of rotation i with respect to base coordinate system
α	Angle used for solution of inverse kinematics problem of 3-DOF manipulator
α_1, α_2	Angles of rotations of the sprockets 1 and 2
α_C	Calculation convergence speed gain of numerical inverse kinematics
$\dot{\alpha}_1,\ \dot{\alpha}_2$	Angular velocities of the sprockets 1 and 2

$\dot{\alpha}_{Ki}$	Angular velocity of i-th wheel
β	Angle used in inverse kinematics of the 3-DOF manipulator (Sect. 4.3)
$\dot{\beta}$	Angular velocity of the robot frame with respect to the instantaneous center of rotation
γ	Angle of slope inclination
γ_P	Proportionality constant for calculation of inverse kinematics
δW	Virtual work
δp	Virtual linear displacements of end-effector
δq	Generalized virtual joint displacement
$\delta \xi$	Generalized virtual end-effector displacement (pose change)
$\delta \phi$	Virtual angular displacements of end-effector
ΔU_{lim}	Limiting error of voltage measurement
ΔI_{lim}	Limiting error of current measurement
ΔI_x	Deformations of the ground
η	Efficiency
θ	Rotation angle of a rotary joint
θ_{f1}, θ_{f2}	Rotation of pedipulator joints 1, 2
$\theta_{r1}, \theta_{r2}, \theta_{r3}$	Rotation of pedipulator joints 1, 2, 3
Θ_i	Generalized forces
ξ	Pose of end-effector
$\xi_E(p_{Ex}, p_{Ey}, ?_{Ez})$	End-effector current pose
$\xi_E^* \ (p_{Ex}^*, \ p_{Ey}^*, \ ?_{Ez}^*)$	End-effector desired pose
ξ_Δ	End-effector pose difference
σ_i	Coefficient of joint type (0 for prismatic joint, 1 for rotational joint)
τ_x	Shear stresses in the soft ground defined by the Coulomb model
$\omega_{n,0}$	Angular velocity with respect to the base coordinate system
$\omega_x, \omega_y, \omega_z$	Angular velocities with respect to axes x, y, z

Chapter 1
Introduction

Mobile robotics is a rapidly developing branch of robotic science. Mobile robots are used to perform tasks as support or substitution of human actions and consequently fill the gap between functionality of stationary robots and the need of mobile solutions. Application of mobile robots for supervision, penetration and identification of hazardous, inaccessible or unknown environments is constantly becoming a common practice and an indispensable part of industrial automation. The term inspection can be interpreted as control, supervision and checking of installations, machinery or indoor and outdoor space. Modern inspection techniques are utilized to substitute human in dangerous tasks and to introduce monitoring of unreachable places. Due to the fact that for each environment requiring inspection, it is recommended to use a dedicated mobile platform, there exist numerous designs of motion units, sensors, power supplies and control systems, dedicated for mobile robots.

In this book, a versatile robotic system for pipeline inspection is discussed. Pipeline inspection is a common application field of mobile robots because monitoring of inaccessible, long and tight pipelines is a very difficult task for human. Design objective of this system is focused on minimization of number of robots, necessary to inspect different types of horizontal and vertical pipelines, even and rough surfaces. An original, adaptable tracked mobile robot with a patented motion unit is presented. The work includes description of design phases, mathematical modeling, simulations and control system development. Prototype construction process and testing procedures are presented and supplemented with laboratory and field experiments.

1.1 Aim and Scope

The aim of this book is to present design process, modeling and control system development for a versatile pipe inspection robot. Motivation of the research is to model, simulate and implement an actively adaptive mobile robot. The original, patented

M. Ciszewski et al., *Modeling and Control of a Tracked Mobile Robot for Pipeline Inspection*, Mechanisms and Machine Science 82, https://doi.org/10.1007/978-3-030-42715-3_1

structure of mobile motion unit for a mobile robot gives possibility to adapt to various sizes and shapes of pipelines, oriented horizontally and vertically. Such robot provides advantageous motion capabilities in contrast to other available solutions and gives prospective for optimization of inspection tasks in different branches of industry. An original mechanical design, based on two track drive modules, assembled to an adaptable chassis, enables utilization of the robot in changing conditions.

Mathematical modeling comprises kinematic and dynamic models of robot motion on flat and inclined surfaces that are useful for control of its position and orientation, also during operation in underwater environments.

Control of the robot adaptation to different shapes, sizes of pipelines and other types of operation environment requires special approach to mathematical modeling. The modeling includes formulation of kinematic equations of motion and derivation of inverse kinematic models of the robot's pedipulators. The most important task that is trajectory calculation and control of motion unit pose, is based on an original algorithm, developed by the author, specifically for the structure of the robot. The algorithm is verified numerically by simulations of the motion unit, prepared in the MATLAB software.

Simulations of mobile robots serve as valuable design resources for development of control strategies in changeable operating conditions. In the scope of this book, advanced multibody simulations are conducted, including motion testing of the robot in different environments. Usage of the original pedipulators control algorithm is presented with on-line operator interaction, resembling real prototype operation. The co-simulation environment consists of the MATLAB/Simulink software for mathematical model processing and hardware interaction, whereas the V-REP application provides simulation of physical phenomena and execution of control commands.

Control system of the robot is developed as a solution that incorporates knowledge from the field of mobile robots and arm-type robots, since the discussed robotic system requires a multidisciplinary approach. Custom electronic control board design is outlined with focus on functionality of particular components and their application in the control system. Development of control software for the on-board controller and high-level user application is presented and implementation on a prototype is shown.

Prototype of the robot was built, basing on the 3D model, developed by the author in the Autodesk Inventor software. The process involved construction and integration of mechanical, electrical, electronic systems with sensors and control systems, implemented in the robot and an operator computer. Laboratory tests of the prototype are described and various experiments are presented. The original control algorithm for trajectory planning of robot pedipulators, developed at the mathematical modeling phase of the project is verified during prototype experiments in different work scenarios and operating conditions. On the basis of numerous experiments, the robot prototype was enhanced to provide augmented functionality in the aspect of motion unit, communication, power supply and control system. Enhanced version of the robotic inspection system is presented with focus on control architecture and practical aspects of industrial applications.

1.2 Book

Aim and scope of the book, presented above allowed to formulate the following book:

It is possible to formulate a mathematical model and implement it in a control system for a mobile robot, equipped with two track drives, to actively adapt and realize motion in horizontal pipes of various shapes and sizes, in vertical pipes and on even surfaces.

1.3 Organization of the Book

Chapter 1

This chapter covers general outline of the book, aim, scope of conducted research and design phases. book of the book is formulated, according to the design objectives. Organization and structure of the work is given.

Chapter 2

In this chapter, robotic inspection of pipelines is presented. Literature research on mobile inspection robots and their applications in various fields is described with relation to the book objective. Next, problems and defects that arise during exploitation of pipelines are outlined, along with available solutions for monitoring and inspection. Afterwards, review of mobile robots for pipeline inspection is given. Their different modes of application are described that lead to identification of a market need for a versatile pipe inspection robot.

Chapter 3

This chapter is focused on mechanical design and 3D modeling of the pipe inspection robot. Firstly, mechanical structure with patented robot chassis, based on active pedipulators adaptation mechanism is presented. Secondly, actuators and inspection equipment are described. Finally, possible operation environments and a complete 3D model of the robot are shown.

Chapter 4

This chapter covers mathematical modeling of the robot. Kinematic and dynamic models of the robot motion, dedicated for even surfaces are presented. Since the research is focused on modeling and trajectory planning for robot pedipulators that allow adaptation of the robot to different pipe shapes and sizes, forward and inverse kinematics are discussed and followed by description of an original trajectory calculation algorithm, developed for control of the closed kinematic chains. Application of the algorithm is verified numerically and visually in MATLAB software.

Chapter 5

In this chapter, robot motion and adaptation co-simulations in MATLAB and V-REP environments are described. The simulations feature various motion scenarios, including pipe adaptation, motion on horizontal surfaces, rough terrain, horizontal and vertical pipes.

Chapter 6

This chapter covers control system design for the robot. Literature research on control and navigation of mobile robots is followed by review of control strategies utilized for arm-type robots. General structure of the robot control system is presented. Development of on-board controller is described and software implementation is discussed with focus on application of the original trajectory planning algorithm.

Chapter 7

In this chapter, building process of the robot prototype is presented. It includes integration of mechanical parts, electronic components, inspection equipment, control system and implementation of software. It is followed by numerous laboratory experiments that prove that the book is satisfied. Finally, description of an enhanced version of the prototype is outlined.

Chapter 8

This chapter contains discussion on results of the research and future work that could be conducted to improve functionality, optimize design and facilitate industrial deployment of the robotic system for pipeline inspection.

Chapter 2
Robotic Inspection of Pipelines

2.1 Mobile Inspection Robots

Mobile robots are devices that can operate in various environments, depending on application. The main classification can be applied, according to the type of realized motion: ground robots, climbing robots, flying robots, swimming and underwater robots. For each type of environment, we can distinguish many locomotion strategies. Thus, for instance, ground robots, may be equipped with wheels, tracks, legs, crawling motion units, or jumping mechanisms. Climbing robots can have wheels, legs, tracks, grippers, magnets, utilize crawling motion or additional mechanisms that allow climbing in a specific environment. Flying robots can resemble man-operated devices such as airplanes, helicopters but can be designed as multicopters or unmanned spacecrafts. Swimming robots can be created similarly to ships or submarines or may be constructed to provide easier remote or autonomous operation and better mobility.

Inspection is one of rapidly developing branches of robotics science. Mobile inspection robots are used to monitor places inaccessible to humans or areas with high level of risk. Inspection robots that belong to other subgroup are utilized in industry, where production processes frequently require monitoring at various stages.

Robots belonging to the first group are based mainly on mobile platforms, suitable for various environments. These may include the following operation types: indoor (monitoring of buildings), outdoor (military on-ground applications, hostile environment surveillance, high-voltage power line monitoring), underwater (in-pipe inspection, underwater construction monitoring) and aerial (factory monitoring—exhaust gases, agricultural, surveillance UAV). Increasing demand of new inspection areas pushes research forward and induces development of novel robot designs.

The second group of inspection robots mainly consists of industrial manipulators with specialized equipment used for monitoring tasks. These operations include vision inspection of production processes (fault detection, vision feedback) and measurements of manufactured parts. Practically, an industrial robot may be equipped

M. Ciszewski et al., *Modeling and Control of a Tracked Mobile Robot for Pipeline Inspection*,
Mechanisms and Machine Science 82, https://doi.org/10.1007/978-3-030-42715-3_2

with any type of inspection equipment for desired assessment and validation scenarios. In the scope of this work, ground mobile robots will be discussed.

Mobile robots for ground operation are usually referred to as Unmanned Ground Vehicles (UGV). The simplest ground mobile robots are wheeled platforms. Indoor 3-wheeled robots that feature two driven wheels and one support wheel that are intended for building monitoring and environment recognition, designed at AGH University are described in [1]. These robots can operate in a group and rely on odometry for determination of position and orientation in a 2D space. A prototype of such robot, with a LIDAR sensor for 2D environment mapping is shown in Fig. 2.1a. The Pioneer 3DX is a commercial design that features similar driving mechanism and can be used for various indoor applications as presented in [2]. It can also realize selected tasks outdoors, owing to large wheels with rubber tires.

Robots with special design of wheels are used where the operation space is more problematic to navigate through. Robots with omni-directional wheels that are holonomic have better maneuverability with respect to non-holonomic robots such as car-like Ackerman steering vehicles or differential drive systems, because they can move in any direction including sideways motion, turning at arbitrarily selected radius or turning in place [3]. The robots with omni-directional wheels usually have three or four independently driven wheels.

Robots with four or more wheels are generally used for rough terrain. The IBIS robot, designed by the PIAP, possesses six independently driven wheels on an adjustable chassis that is suitable for operations in difficult and diverse terrains including sand, snow, boggy land or rocks (Fig. 2.1b). The robot can be used for surveillance, rescue missions and military operations [4].

Operation in extraterrestrial space requires special designs of drive mechanisms that provide harsh terrain passing abilities and allow for optimization of energy consumption. Mobile robots constructed for operation in space frequently have wheeled designs with rocker-boogie suspension systems. These systems may consist of 4-wheel or 6-wheel platforms and different mounting arrangements. The Curiosity Rover, developed by NASA that is successfully operated on Mars features a 6-wheel rocker-bogie suspension with two 3-wheel modules mounted on a central leveraging mechanism [5]. The rover is dedicated to inspect, document and analyze planetary environment. The rover for ExoMars mission, developed by European Space Agency has three independently mounted modules equipped with two wheels, mounted on load-averaging linkages. This construction provides uniform load distribution over the surface with usage of passive mechanisms [6, 7]. Additionally, it should be noted that the wheels are independently driven and four wheels have adjustable turning angle to provide maximum mobility in demanding terrain. The rover can be used for exploration of unknown terrain and features drilling device for soil and rock sample collection [8]. A lightweight version of a planetary drilling and sample collection system was developed at the AGH University and the Space Research Centre of Polish Academy of Sciences (CBK PAN). The system consists of a 4-wheeled rover with 2-wheel rocker modules, mounted to a central differential mechanism [9]. It is equipped with a drilling module for rock sample extraction and storage. The rover construction provides uniform load distribution and roll angle compensation for stabilization

Fig. 2.1 Wheeled mobile robots: **a** indoor mobile robot—KRIM, AGH; **b** IBIS robot—PIAP [4]; **c** Ultralight Mobile Drilling System—AGH, CBK PAN

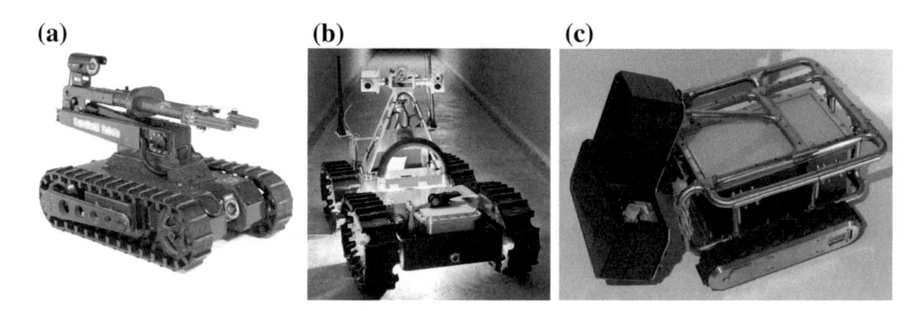

Fig. 2.2 Tracked mobile robots: **a** SDR LT2/F Bulldog robot [11]; **b** SANDIA Gemini Scout robot [13]; AGH University Underwater tank inspection robot [14]

of the drilling subsystem [10]. Prototype of the Ultralight Mobile Drilling System is shown in Fig. 2.1c.

Mobile robots equipped with track drive can be used in tough environments. Tracks generally provide good traction, large robot-ground contact area and thus optimal load distribution with ability to move on soft terrain. Nevertheless, track drives require higher energetic expenses for the travelled distance, compared to wheel drives. There exists wide variety of mobile robots with track drives. An inspection robot equipped with two track drives and a 4-axis manipulator was developed by Super Droid Robots Inc. The robot, called LT2/F Bulldog, is able to negotiate obstacles and climb stairs. It is shown in Fig. 2.2a. Equipment of the robot makes it useful for surveillance of dangerous areas, remote manipulation, opening doors and remote video inspection [11]. Inspection and rescue operations in underground mines are preferably assisted by robots. The Gemini Scout robot by SANDIA is a device with four track modules that give very good ability to traverse demanding terrain such as flooded land or stairs (Fig. 2.2b). The robot is equipped with gas monitoring sensors and underground communication system. Ground mobile robots can also be utilized for underwater operation. A tracked robot for tank inspection and cleaning was designed at AGH University. It can perform 3D mapping of underwater space such as water and other liquid storage tanks by application of a sonar [12]. Additionally, the robot can be used to remove sediments from tank floor with usage of an optionally mounted suction pump. The construction features watertight track drive modules with integrated motors and gear transmissions, assembled to an adjustable housing (Fig. 2.2c).

Fig. 2.3 Other types of ground mobile robots: **a** Toshiba tetrapod robot [16]; **b** NASA RoboSimian robot [17]; **c** Boston Dynamix RHex robot [18]

Hybrid constructions that include wheels and multiple, independently controlled track drives are also utilized for inspection. The PIAP Scout robot possesses four track drives among which two front ones have adjustable inclination angles that provide better ability to negotiate large obstacles [15]. The chassis is augmented with 4-wheels that further improve mobility of the robot.

There also exist inspection robots with other types of locomotion. For the most demanding environments, use of legged, walking robots is preferable, due to the fact that these robots can execute much more complicated movements. Toshiba corporation produces an inspection robot that features four legs and is capable of passing high obstacles and stairs (Fig. 2.3a). This tetrapod robot is able to operate 300 days in high radiation environment to help in cleaning tasks after nuclear disasters [16]. A multi-task walking robot, shown in Fig. 2.3b, was developed by NASA that uses four 7-DOF articulated arms. The high dexterity structure can be also used for manipulation tasks, since its limbs have integrated grippers. The robot is equipped with sensors, scanners and additional wheels for faster locomotion on flat surfaces [17]. Another approach to design of walking robots is used by the RHex robot developed by Boston Dynamics. The robot features six curved legs that are mounted on independently driven rotary actuators. It can operate in mud (Fig. 2.3c), sand, snow or surfaces inclined by up to 30° [18]. Special algorithms provide jumping over obstacles, inverting and climbing. The robot is equipped with visible light and IR light cameras for day and night outdoor operation as well as surveillance of confined spaces.

It can be concluded that numerous application fields of mobile inspection robots induce development of wide variety of designs. Each structure implies creation of a task-specific control system that integrates mathematical models, sensor information, properly selected actuators and controllers to realize motion in demanding environments, faced by mobile inspection robots.

2.2 Inspection of Pipelines

Inspection of pipelines has several objectives. After building a new pipeline, it is required by regulations that the pipe must be inspected by a device, equipped with a camera. In addition, inclination of the pipe has to be precisely measured. With the initial inspection, engineering errors can be found and eliminated before exploitation of an industrial facility. Other uses of inspection techniques include periodical checks of pipe condition or documentation of old pipes.

2.2.1 Pipe Defects

Depending on pipe material, different defects can be found in pipelines. In Fig. 2.4, defects that are likely to develop in pipes are presented. Figure 2.4a depicts coating defect that appeared after long exploitation time, Fig. 2.4b shows corrosion of a steel pipe in a preliminary phase, whereas Fig. 2.4c presents radial deformation of pipe wall caused by excessive ground pressure.

Defects that appear in pipes may be much more drastic and lead to malfunction or blockage of fluid flow. Figure 2.5 depicts blockage by corrosion that reduces cross section of pipe, formation of sediments that solidify on the bottom of pipe and penetration of pipe wall by roots that may lead to leaks and breakdowns of pipe walls.

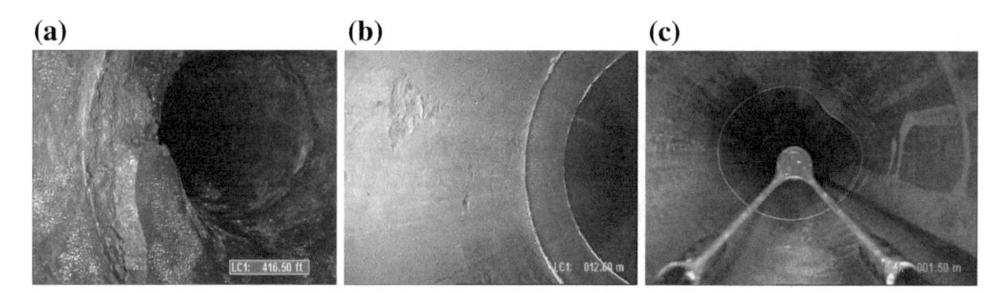

Fig. 2.4 Pipe defects [19]: **a** coating defect of a cement water line; **b** corroded steel pipe; **c** gas pipeline deformation

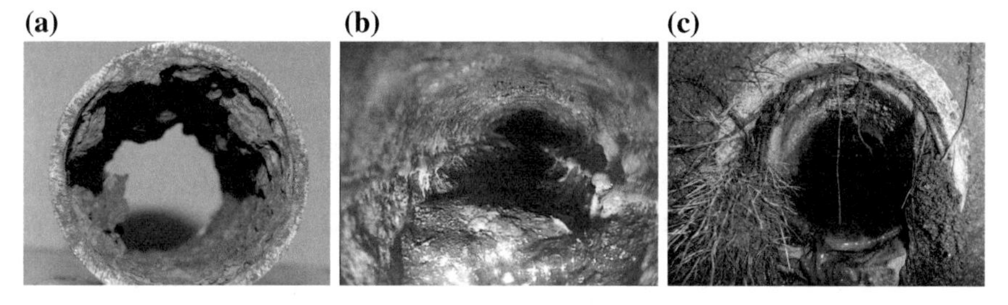

Fig. 2.5 Pipe blockage: **a** a metal pipe microbiologically influenced corrosion [20]; **b** formation of sediments [19]; **c** pipe walls penetrated by roots [21]

To prevent leaks and excessive damage of pipelines, inspection methods are used to avoid deterioration of pipe structure and plan maintenance procedures. The majority of inspection methods require long distance operation in hardly accessible environments, thus, mobile inspection robots are utilized for these tasks.

2.2.2 Pipe Inspection Methods

Extensive structure and diversity of industrial facilities require application of various inspection methods. Among the most useful are non-invasive techniques that do not influence material structure of the investigated element. Inspection of inside of pipelines can be realized without the necessity to exclude it from exploitation [22]. Systematic control of pipelines should be included in a proper maintenance strategy but inspection is also used for new fragments of pipes after construction. This approach, taken as standard in the last few years, provides easy control of assembly quality and gives possibility to eliminate faults before exploitation phase. The most common problems that arise in pipelines under exploitation include: excessive sediment formation, damage of pipe walls, intrusion of foreign objects, corrosion, pipe profile longitudinal and radial deformations. To detect these undesirable phenomena that may appear in unreachable pipe segments, diverse Non-Destructive Testing (NDT) inspection techniques are utilized. Non-Destructive Evaluation (NDE) is a term for more quantitative measurements that can be used interchangeably with NDT. A method in which a defect is located and also quantified by size, location, shape, orientation or change of material properties would be referred to as NDE [23].

The most common inspection method is a visual, optical investigation that includes mounting of a Closed-Circuit TeleVision (CCTV) camera on a mobile platform that can be remotely operated with video capture and analysis. A CCTV camera system includes low light sensitive sensor that provides real-time video Images/image transmission to an operator. Sensors utilized in these cameras are manufactured using Charge Coupled Device (CCD) or Complementary Metal Oxide Semiconductor (CMOS) technologies. Generally, CCD sensors exhibit better light sensitivity, but constant development of CMOS sensors that are faster and more energetically efficient leads to wider usage of this technology in practice [24]. Moreover, different types of inspection camera heads are utilized. The simplest ones feature constant focal length sensors with wide angle of view (Fig. 2.6a). These cameras provide Images/image of the entire pipe circumference and are usually mounted in the center of an inspection robot. More advanced camera heads feature a Pan-Tilt-Zoom (PTZ) mechanism that allows the camera to rotate (tilt) by 360° with respect to longitudinal axis and rotate (pan) to the side. Additionally, some heads with zoom, as the one shown in Fig. 2.6b can provide detailed Images/image of even very small pipe wall damages. Extensive ranges of cameras are offered by pipeline inspection companies such as IBAK [25], IPEK [26], KANRES [27]. The camera heads differ in size, resolution and illumination characteristics. Some cameras provide additional position adjustments that allow inspection of pipe branches (Fig. 2.6c). It should be

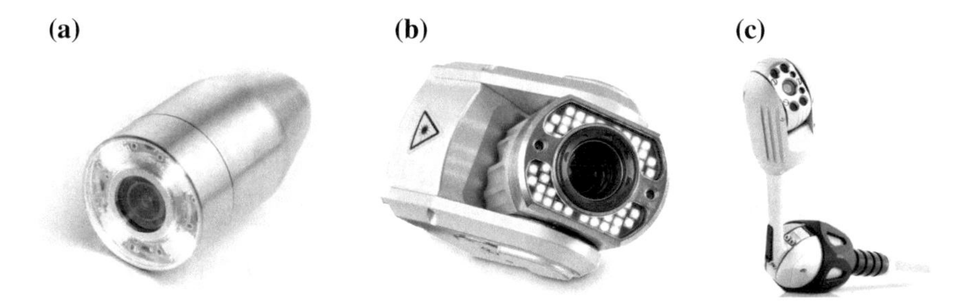

Fig. 2.6 Pipe inspection CCTV cameras: **a** Inuktun Mini Crystal Cam constant focus camera [28]; **b** IPEK RCX 90 Pan-Tilt-Zoom camera [26]; **c** IBAK RapidView Pan-Tilt-Zoom camera with an adjustable arm for pipe branches [25]

noted that a complete CCTV inspection system is normally equipped with a screen, a video recorder and a control panel for the operator that allows adjustment of mobile robot speed, provides manipulation of camera and displays video signal.

Another type of pipe inspection is based on ultrasonic waves propagation in solid bodies. Ultrasonic testing method uses high-frequency waves (0.1÷50 MHz), transmitted into a material to detect imperfections or to locate structural changes. The most commonly used ultrasonic testing technique is the pulse echo, where waves are introduced into an investigated object by a specialized head and reflections (echoes) from internal imperfections or the part's geometrical surfaces are returned to a receiver [29]. In this method one transducer is utilized that serves as transmitter and receiver of ultrasonic waves. The second method is attenuation or through-transmission method, when separate transmitter and receiver are used. Defects are localized by the attenuation of ultrasonic waves, transmitted by material. In modern ultrasonic testing equipment, a dedicated diagnostic machine can display information about distance of a defect from ultrasonic transducer and size of the imperfection (Fig. 2.7a). These methods usually require an addition medium such as oil or water to allow transmission of ultrasonic waves to the investigated material, but contactless Electromagnetic Acoustic Transducers can be also utilized [30]. The ultrasonic testing method can be utilized to evaluate pipe corrosion and imperfections, however specific set of transducers have to be applied for each material type. For long section of pipes, the ultrasonic testing is used on large diameter pipe inspection devices "pigs", propelled by fluid flow inside of pipe, equipped with sets of sensors, placed on their entire circumference [31]. Alternative ultrasonic testing method of high accuracy and sensitivity for detection of defects in welds is called Time-Of-Flight Diffraction (TOFD). The method relies on measurement of amplitude of reflected signal from different surfaces. When a crack is present in the weld, there is a diffraction of the ultrasonic wave that propagates from defect edges. Measured time of the pulse gives information about size and location of an imperfection. Ultrasonic testing may be used as a robotized method for weld structure evaluation [32]. Development of new ultrasonic inspection methods that can be adapted to various materials is conducted by research teams [33] and non-contact methods are being constantly developed [34].

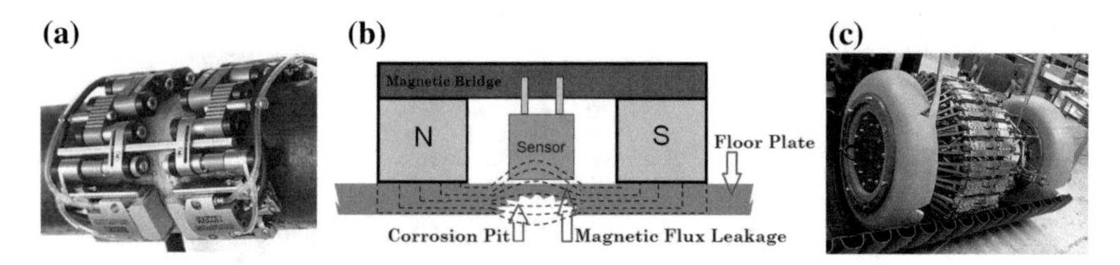

Fig. 2.7 NDT inspection methods: **a** Olympus Cobra outside pipe scanner [35]; **b** Magnetic Flux Leakage method—scheme [36]; **c** "Pig" probe with an array of MFL sensors [37]

Radiography is a method that involves usage of penetrating gamma or X-radiation in materials for detection of structural imperfections. An X-ray generator or radioactive isotope is used as the source of radiation. For this type of inspection, access from both sides of pipe is required, because radiation is directed through a part and onto a film or another detector. The most common features, examined by this method using difference of radiograph illumination are metal pipe welds [23].

Magnetic field can also be used for inspection of pipelines. The Magnetic Flux Leakage (MFL) method utilizes high strength magnets to temporarily magnetize steel structure. If a defect is present in material, magnetic field is distorted and the "leakage" of magnetic flux can be detected by sensors, providing information about corrosion, pitting or structural discontinuity (Fig. 2.7a). This method is widely used for large-diameter pipes where an array of sensors is located around a "pig" inspection tool, as depicted in Fig. 2.7c. The MFL method can also be utilized for small diameter ferromagnetic pipes.

There exist various electromagnetic inspection methods for metal pipes, yet eddy current testing is the most popular one [23]. It involves generation of electrical currents (eddy currents) in a conductive material by changing magnetic field. The intensity of magnetic field, amplitude and phase shift of voltage and current are measured and analyzed to provide information about material defects that cause interruptions in the flow of these currents. The method is widely used for inspection of pipe wall thickness, however very accurate, it suffers from limited penetration depth for ferromagnetic materials. Application of Remote Field Eddy Current Testing method increases penetration depth in ferromagnetic materials [22].

Recent development of laser technology lead to introduction of pipe profiling techniques with usage of a device, integrated in the camera or a dedicated laser scanner. Application of a laser scanner gives precise measurements of pipe diameter, deflection and deformation [38]. Additional usage of integrated CCTV camera extends geometric information with visual inspection data (Fig. 2.8a). As a result of inspection, a 3D model of pipe can be obtained that allows analysis of pipe ovality at each section. Additionally, if such system is equipped with odometric measurement and an inclination sensor or an Inertial Measurement Unit (IMU), it is possible to build a 3D model that provides information about longitudinal pipe profile (Fig. 2.8b).

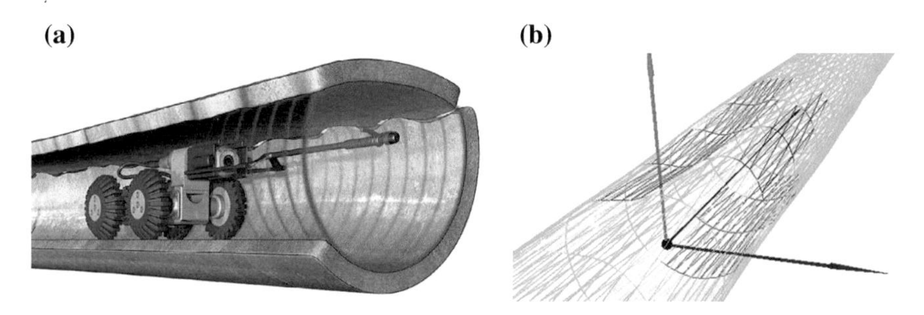

Fig. 2.8 Laser profiling of pipes: **a** robotic laser profiling [39]; **b** 3D model of a pipe [38]

The analysis of pipeline inspection methods shows that the most versatile technique that can be applied to pipes made of different materials with changing structure is video inspection by a CCTV camera. It is common for general pipe inspection tasks in water, sewage, oil and gas industries. Video inspection can be supplemented with data gathered from other sensors to improve monitoring quality.

2.3 Mobile Robots for Pipeline Inspection

Pipelines generally require two types of inspection procedures. Since the majority of pipe networks are located underground, generally the inner surface has to be investigated in order to assess technical condition of a pipeline. In some cases, pipes that transport gases or liquids under pressure may be mounted vertically or arranged as overpasses with multiple bends, frequently present in factories. This kind of pipes require different monitoring approach that can be realized in two ways. Inspection robots for outer pipe surfaces can be used or a particular section of pipe may be inspected by an in-pipe robot with driving mechanism, adapted to travel vertically and optionally pass connectors and bends.

Numerous research teams work on development of pipe inspection robots that meet different needs in terms of structure, mobility, size and adaptation to work environment. Pipe inspection robots are also produced by companies that offer products for customers from different branches of engineering industries.

In-pipe robots have wide range of applications and are one of the most popular types of mobile robots used in industry. Depending on work environment, different locomotion units are employed such as wheels, tracks, legs, crawling systems or passive mechanisms of pig-type robots that allow motion by fluid flow [40]. Wheeled robots are the most common for inspection of horizontal pipes, however they suffer from lower mobility in pipes with complicated layout, including bends, reducers and branches. Tracks provide high traction and enhanced obstacle negotiation performance. Legs provide motion in complicated pipe layouts, but are not widely used for inspection tasks due to problems with control systems, required to ensure steady camera recording. Crawling locomotion can be used for complicated pipe networks with considerable shape changes but suffer from the analogous stability problems as walking robots. Some uses of in-pipe robots require adaptation to pipe diameters and shapes as well as passing through bends, branches or vertical segments. Hybrid drive systems have been developed by several researchers that provide drive adaptation and motion in vertical pipes [41].

2.3.1 Outside of Pipe Inspection

Inspection robots that move on the outside surface of pipes are not frequently used in industry and are still in the research phase. The 3DClimber robot, presented in [42] is designed to move on the outside surface of pipes. It can be classified as a 4-DOF articulated arm with dedicated V-shaped grippers that allow passing through bends and T-junctions. The robot is equipped with force sensors and rangefinders to detect changes in pipe structure. A robotic system described in [43] consists of modules that can be assembled to create an out-pipe climbing robot that uses magnetic adhesion in an adaptive gripper. However, the prototype is not equipped with sensors for inspection. Another concept of an out-pipe inspection robot was developed, based on omnidirectional wheels [42]. The robot is able to move on flat or curved ferromagnetic surfaces and exhibits high level of maneuverability. In recent year the increasing popularity and accessibility of Unmanned Aerial Vehicles (UAV) have led to development of this kind of robots for outside pipe inspection [44].

2.3.2 In-Pipe Inspection Robots

Literature research shows various concepts of in-pipe robots. Wheeled inspection platforms based on modular design that feature segments with wheeled legs on pantograph mechanisms for adaptation to pipe dimensions are presented by [40]. The proposed robots MRINSPECT-III and MRINSPECT-IV are suitable for ⌀200 mm and ⌀85÷⌀109 mm pipes and feature CCD cameras with lighting. They have extensive steering capabilities that allow operation in pipes with any orientation and complex branches. A wheeled mobile robot for inspection and pipe mapping was presented by [45]. The robot is designed for movement in horizontal pipes and utilizes a forward-facing fisheye camera. This setup allows to perform visual odometry and build pipe models with high resolution. An inspection robot that is capable of removing obstacles and cleaning of pipes was described by [46]. The authors also tested a wireless control system for the robot. An alternative design of a tracked robot was presented in [47]. They proposed a platform with a cylindrical track drive, called Omni-Track. In addition to conventional forward and backward motion, the tracks are able to move sideways by roll mechanism. As the article presents, this design increases contact area when moving in pipes with various diameters, both inside and on outside surfaces. A concept of three-track vertical configuration was proposed by the authors.

Market research for inspection robots showed several interesting solutions available for industrial customers. iPEK company produces wheeled inspection vehicle series, named the ROVVER, for pipes with diameters ∅100÷300 mm, ∅150÷760 mm and ∅230÷1520 mm [48]. These robots have modular design, with replaceable wheels that can be mounted depending on work environment. They are rated for 10 m underwater operation and can be used for inspection of various types of horizontal pipes. ABE Complex Technologies offers four different wheeled inspection robots that together provide inspection of pipe with diameters in range ∅130÷1800 mm [49]. They are equipped with Pan-Tilt-Zoom (PTZ) cameras and exchangeable wheels (Fig. 2.9a). Inuktun produces wide range of tracked inspection robots. Versatrax models are available in three different sizes for minimal pipe diameters: ∅100, ∅150 and ∅300 mm [28]. Their main components are individually operated tracks of different sizes. Each robot chassis allows manual adjustment of the driving mechanism to suit particular task. They can be used for sewer and storm drains, air ducts, tanks, oil or gas pipelines pulp and in paper industry. The largest of the robots in range, the Inuktun Versatrax300 is depicted in Fig. 2.9b. Another design by Inuktun, the VGTV vehicle has variable geometry track that can change position of mounted camera and transform its shape from straight to triangular, vertically oriented, that is well suitable for muddy conditions [28]. All Inuktun products are equipped with CCTV cameras with lights and are rated for operation 30 m underwater. Autonomous Solutions company produces the Chaos robot, equipped with four tracks with adjustable inclination angle that can accommodate to encountered terrain. It can be utilized for inspection of pipes larger than ∅800 mm in diameter and it is not waterproof [50]. However, wireless operation is a great advantage of this robot. The Solo robot by RedZone is a wireless, autonomous, tracked device that can be used in horizontal ∅200÷300 mm pipes [51]. It features slim tracks that cover majority of robot base and provide high friction. Another RedZone product is the Responder robot that has two massive tracks and can incorporate different instrumentation, even a 3D laser scanner. It is intended for pipes with diameter higher than ∅800 mm and can operate up to 45 m underwater [52]. CUES company offers tracked inspection robots for pipes variable from ∅150 to ∅700 mm. Locomotion relies on narrow tracks made of large segments, shown in Fig. 2.10a. They are equipped with cameras and lights. Height of the robots can be adjusted by extenders to provide appropriate camera position [53].

A modular snake robot, representing crawling locomotion group, described in [55] was specifically designed to pass horizontal and vertical waves through its body to move in three-dimensional spaces, like pipelines or rubble. The robot measures 50 mm in diameter and its length exceeds 900 mm. Its body consists of 16 joints, arranged perpendicularly to each other that allow motion with usage of variety of gaits in an analogous way to a biological Sidewinder snake, as depicted in Fig. 2.10b. The robot's performance suffers from insufficient stability for continuous video inspection.

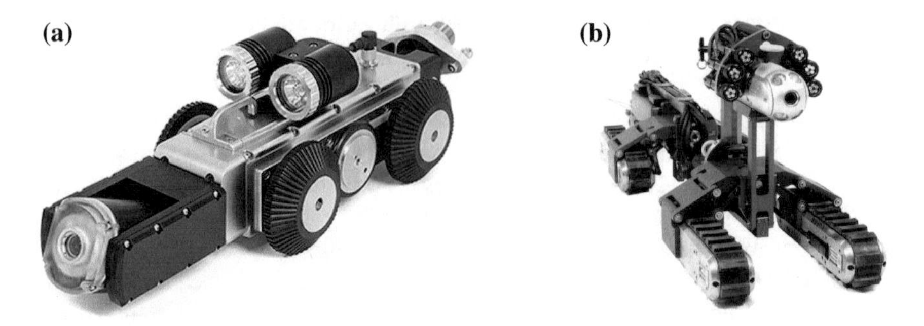

Fig. 2.9 Horizontal in-pipe inspection robots: **a** ABE CT-400 wheeled inspection robot with a PTZ camera [49]; **b** Inuktun Vesatrax 300 tracked robot [54]

Fig. 2.10 Horizontal in-pipe inspection robots: **a** CUES Ultra Shorty 21 [53]; **b** a snake-like robot with 16 links [55]

2.3.3 *Vertical In-Pipe Inspection*

Nowadays, pipe inspection robots intended for operation in vertical segments are becoming more popular. The Versatrax Vertical robot is a three-track version of the Inuktun Versatrax series for vertical pipes, initially developed for dry pipe inspection [56], then accustomed for water environment [57]. The robot features pantograph track adjustment mechanism and can be used in vertical pipes with diameters in range ∅200÷300 mm. It is shown in Fig. 2.11a. The three-track robot, presented in [58] is designed for inspection of large diameter gas pipelines that can be oriented vertically. It features an active pipe-diameter adaptive mechanism based on pantographs that provides adjustment of track extension force control and motion in pipes with diameters ranging ∅400÷650 mm. Another solution is provided by Neovision company that offers a tracked robot, Jetty, intended for inspection and cleaning of vertical ducts [59]. It features six track drive modules, mounted on pantograph mechanisms (Fig. 2.11b) and provides inspection of circular and rectangular ducts with diameters from ∅400 to ∅1300 mm in extended arm version. Motion inside of ferromagnetic vertical pipes can be also realized by magnetic crawlers. The robot presented in [60] features one-piece magnet, but two-piece magnet models and can be used as well

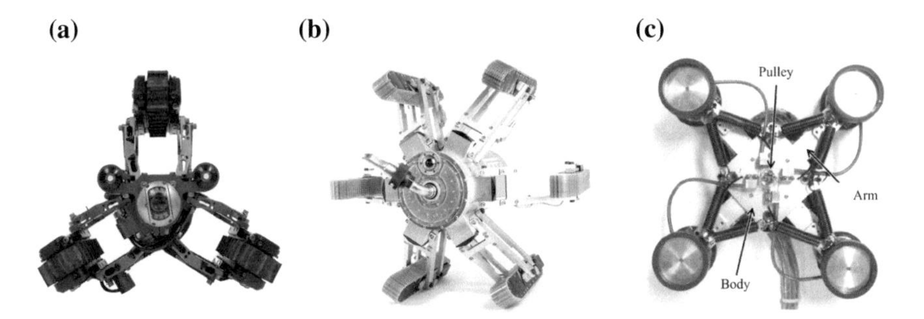

Fig. 2.11 Vertical in-pipe inspection robots: **a** Inuktun Versatrax Vertical [59]; **b** Neovision Jetty robot [59]; **c** the AQAM robot [62]

inside of vertical pipes or on flat vertical ferromagnetic surfaces. However, operation of this solution is limited to surfaces made of specific material. Similar approach was used by Inuktun company by their Versatrax 100 MicroMag robot [28]. A tracked robot designed to move in distant sea-water pipelines with horizontal and inclined linear segments, variable from ∅600 to ∅800 mm in diameter was described in [61]. The driving module has three pantograph-type links, equipped with tracks, spaced in 120°. This design makes it possible to realize the adaptation to pipe diameter and the adjustment of wall-pressing force. Nevertheless, it does not provide motion in rectangular pipes or horizontal surface.

Wheeled robots for vertical pipes, developed by research teams have different configurations. The robot presented in [62] has four pairs of wheels, mounted on the Adaptable Quad Arm Mechanism (AQAM) that can change its shape to drive through pipe branches (Fig. 2.11c). Wheel modules extension that provides traction force is realized towards two opposite pipe walls. In addition, it features adjustable angle of wheel sideways rotation that gives possibility to move in spiral trajectories. A robot intended for operation in round pipes with constant diameter and shape was described in [29]. The robot chassis features six wheels that are assembled in two sets of three wheels, spaced in 120°. One segment is used as a stator with wheels oriented in the direction of pipe, whilst the other one with inclined wheels propels the robot by rotation about robot's longitudinal axis. Adaptation to pipe diameter can be realized by change of rotor and stator arms. The same concept was described in [63] but was developed only up to initial prototype stage. The robot has a stator and a rotor with inclined wheels that allows propelling forwards and backwards by realizing helical trajectories with respect to pipe walls. Another wheeled robot for horizontal and vertical pipes was proposed in [63]. It is intended to operate in pipes with constant radius. Its construction allows to use only one actuator to pass through pipeline elbows.

Snake-like climbing robots are also developed for vertical pipe inspection. A design described in [64] gives possibility to move in variably shaped vertical pipes by application of the "sinusoidal wave drive" principle. The 13-segment structure, equipped with a camera propels the robot by clinging to pipe walls and provides very high level of flexibility in passing bends, T-shapes and changes of diameter. It is however very unstable in retaining camera position and radial extension force. Thus, vertical operation is not stable and this kind of robot may be susceptible to accidental falls.

2.4 Summary

Mobile inspection robots are mechatronic devices commonly used in many application fields. They can be utilized for different tasks; thus, their structural and control architectures differ significantly. Changing environments of operation induce numerous challenges in modeling, design and control of mobile robots.

Exploitation of pipelines and development of new pipe networks require diverse maintenance techniques and inspection methods to ensure appropriate technical condition, unobstructed flow of transported fluids and leak prevention. Among numerous NDT testing methods, the most popular technique is visual inspection with usage of a CCTV camera, mounted on a mobile robot.

Numerous robotics solutions for pipeline inspection are available. In horizontal pipes, wheels provide the least rolling resistance and are energy efficient, however their small contact area may not be sufficient for some uneven surfaces. Crawling locomotion has speed limitations and especially upper limits of pipe or duct dimensions are its major drawbacks. As presented by market research, numerous solutions utilizing track drive have been developed. Tracks provide proper obstacle passing capabilities and considerably large contact area that gives satisfactory traction in changing environment.

The reviewed robots designed for vertical pipes cannot be used in general for non-circular cross section pipes or flat and uneven surfaces. Drive adaptation capabilities are also very limited in most solutions. Whilst some robots provide manual drive adjustments, they do not allow motion in diverse and changing environment. A need for an actively adaptable robot that could be accustomed to various types of pipe inspection tasks was identified, in accordance with the literature research.

The aim of this book is to present design, modeling, control and prototype development of a robot that would be able to move in pipes and ducts with round and rectangular cross-section that are horizontal or vertical, with bends and interconnections. Versatile chassis of the robot would make it easy to deploy in other inspection tasks, thus minimizing number of necessary devices, required for monitoring of industrial facilities.

References

1. Buratowski T, Dabrowski B, Uhl T, Banaszkiewicz M. The precise odometry navigation for the group of robots. Schedae Informaticae. 2010;19:99–111.
2. Francis SLX, Anavatti SG, Garratt M. Dynamic model of autonomous ground vehicle for the path planning module. In: ICARA 2011—Proceedings of the 5th international conference on automation, robotics and applications;2011, p. 73–77.
3. Doroftei I, Grosu V, Spinu V. Design and control of an omni-directional mobile robot in novel algorithms and techniques in telecommunications, automation and industrial electronics. Dordrecht: Springer Netherlands; 2008. p. 105–110
4. Przemysłowy Instytut Automatyki i Pomiarów PIAP. Robot Mobilny IBIS. 2014. http://antiterrorism.eu/wp-content/uploads/ibis-pl.pdf. Accessed 16 May 2016.
5. NASA. Mars Science Laboratory (MSL). http://nssdc.gsfc.nasa.gov/nmc/spacecraftDisplay.do?id=2011-070A. Accessed 12 Apr 2015.
6. Van Winnendael M, Baglioni P, Vago J. Development of the ESA ExoMars rover. In: Proceedings 8th international symposium on artificial intelligence, robotics and automation space;2005, p. 5–8.
7. Kucherenko V, Bogatchev A, Van Winnendael M. Chassis concepts for the ExoMars rover. In: The 8th ESA workshop on advanced space technologies for robotics and automation (ASTRA'04) (2004).
8. Martowicz A, Ciszewski M, Buratowski T, Gallina A, Rosiek M, Seweryn K, Teper W, Zwierzyński AJ, Uhl T. Mechatronic approach in application to solution of research and design problems. Mechatronics. 2016; 36:1–17.
9. Ciszewski M, Buratowski T, Uhl T, Giergiel M, Seweryn K, Teper W, Zwierzynski AJ. Ultralight mobile drilling system-design and analyses of a robotic platform intended for terrestrial and space applications. In: Robot motion and control (RoMoCo);2015. p. 84–90.
10. Ciszewski M, Teper W, Buratowski T, Uhl T, Gallina A, Seweryn K. Design of an ultralight mobile platform for a drilling system. 2015 IFToMM World Congress Proceedings;2015.
11. SuperDroid Robots Inc. SDR Tactical Robots. 2016. http://www.sdrtactical.com/. Accessed 17 May 2016
12. Buratowski T, Ciszewski M, Giergiel M, Siatrak M, Wacławski M. Mechatronic approach in inspection of water supply networks;2015 p. 317.
13. Green J. Mine rescue robots requirements Outcomes from an industry workshop. In: 6th robotics and mechatronics conference (RobMech);2013, p. 111–116.
14. Ciszewski M, Giergiel M, Kudriashov A, Małka P. Modelowanie i analiza modalna ramy mobilnego robota inspekcyjnego. In: Modelowanie Inżynierskie. 2015; 23.54:20–25.
15. Przemysłowy Przemysłowy Instytut Automatyki i Pomiarów PIAP. Robot mobilny Scout. 2014. http://antiterrorism.eu/wp-content/uploads/piap-scout-pl.pdf. Accessed 16 May 2016.
16. Hixson L. New four legged robot developed by Toshiba to search inside Fukushima reactors. Enformable.com, 21 Nov 2012. http://enformable.com/2012/11/new-four-legged-robot-developed-by-toshiba-to-search-inside-fukushima-reactors/.
17. Hebert P, Bajracharya M, Ma J, Hudson N, Aydemir A, Reid J, Bergh C, Borders J, Frost M, Hagman M, Leichty J, Backes P, Kennedy B, Karplus P, Satzinger B, Byl K, Shankar K, Burdick J. Mobile manipulation and mobility as manipulation- design and algorithms of robosimian. J Field Robot. 2015;32(2):255–74.
18. Boston Dynamics. RHex All-Terrain Robot. 2016. http://www.bostondynamics.com/robot_rhex.html. Accessed 17 May 2016.
19. Maverick Inspection LTD. Maverick—Pipelines. 2016. http://www.maverickinspection.com/services/remote-video-inspection/rvi-video-imagery-gallery/pipelines-compressors/. Accessed 04 Apr 2016.
20. LEGEND Technical Services Inc. Microbiologically Influenced Corrosion (MIC). http://www.legend-group.com/node/170. Accessed 07 Apr 2016.
21. Ontario Septic Tank Information. Roots Around Inlet Pipe. http://ontariosептictank.ca/olympus-digital-camera-55/. Accessed 07 Apr 2016.

22. Kurc K. Mechatronika w projektowaniu robota. Rzeszów: Oficyna Wydawnicza Politechniki Rzeszowskiej, 2010. Chapter 2. Robotic Inspection of Pipelines 20.
23. NDT Resource Center. About NDT. https://www.nde-ed.org/AboutNDT/aboutndt.htm. Accessed 15 Apr 2015.
24. Janesick JR, Elliott T, Andrews J, Tower J, Pinter J. Fundamental performance differences of CMOS and CCD imagers: part V. In: Proceedings SPIE 8659, sensors, cameras, and systems for industrial and scientific applications XIV (2013), p. 865–902.
25. IBAK Helmut Hunger GmbH & Co. KG. Cameras—Technological masterpieces & practical designed. http://www.ibak.de/en/produkte/ibak_show/frontendshow/category/kameras/. Accessed 17 May 2015.
26. IPEK International GmbH. iPEK pipeline inspection. 2016. https://www.ipek.at/. Accessed 05 Oct 2015
27. KANRES Sp. z o.o. Kamery do kanalizacji. http://www.kanres.pl/produkty/kamery%20do%20kanalizacji. Accessed 12 Apr 2015.
28. Inuktun Services Ltd. Inuktun crawler vehicles. 2015. http://www.inuktun.com/crawler-vehicles. Accessed 25 Oct 2015.
29. Giergiel J, Kurc K, Giergiel M. Mechatroniczne projektowanie robotów inspekcyjnych. OficynaWydawnicza Politechniki Rzeszowskiej; 2010.
30. Hirao M, Ogi H. EMATs for science and industry—noncontacting ultrasonic measurements. Springer US, 2003. Chapter 2. Robotic Inspection Of Pipelines 19.
31. Holstein P, Andreas T, Ulf B. Ultrasonic pig detection at pipelines. PPSA Seminar; 2010.
32. Dobie G, Galbraith W, Macleod C, Summan R, Pierce G, Gachagan A. Automatic ultrasonic robotic array. In: IEEE International Ultrasonics symposium, IUS; 2013. p. 1861–1864.
33. Dziedziech K, Pieczonka L, Kijanka P, Staszewski WJ. Enhanced nonlinear crack-wave interactions for structural damage detection based on guided ultrasonic waves. In: Structural control and health monitoring 23.8;2016. p. 1108–1120.
34. Kaczmarek M, Piwakowski B, Drelich R. Noncontact ultrasonic nondestructive techniques: state of the art and their use in civil engineering. J Infrast Syst. 2016; 23.1.
35. Olympus Corporation. Olympus COBRA scanner. 2016. http://www.olympus-ims.com/en/scanners/cobra/. Accessed 02 Apr 2016.
36. MFE Enterprises Inc. What is MFL? 2016. http://www.mfeenterprises.com/what-is-mfl. Accessed 05 Apr 2016
37. Pure Technologies Ltd. PureMFL Magnetic Flux Leakage Inline Inspection (ILI). 2016. https://www.puretechltd.com/technologies-brands/pure-mfl. Accessed 03 Apr 2016.
38. Rausch Electronics USA LLC. Rausch M-Series Laser Pipe Profiling System. 2016. https://rauschusa.com/products/laserprofile. Accessed 19 May 2016.
39. Mini-Cam Ltd. Mini-Cam ProLaserTM Profiling. 2016. http://www.minicam.co.uk/laser-profiling/. Accessed 10 May 2016
40. Choi HR, Roh S. In-pipe robot with active steering capability for moving inside of pipelines in Bioinspiration and robotics walking and climbing robots. InTech. 2007.
41. Roslin NS, Anuar A, Jalal MFA, Sahari KSM. A review: hybrid locomotion of in-pipe inspection robot. In: Procedia engineering; 2012, p. 1456–1462. Chapter 2. Robotic inspection of pipelines 21.
42. Tavakoli M, Lopes P, Sgrigna L, Viegas C. Motion control of an omnidirectional climbing robot based on dead reckoning method. Mechatronics. 2015;30:94–106.
43. Maempel J, Koch T, Koehring S, Obermaier A, Witte H. Concept of a modular climbing robot. In: 2009 IEEE symposium on industrial electronics & applications 2; 2009. p. 789–794.
44. Ascending Technologies GmbH. UAV Inspection, Monitoring of Industrial Assets, Oil & Gas. 2016. http://www.asctec.de/en/uav-uas-drone-applications/aerial-inspection-aerial-monitoring/#pane-0-1. Accessed 26 Apr 2016.
45. Hansen P, Alismail H, Rer P, Browning B. Visual mapping for natural gas pipe inspection. Int J Robot Res. 2015;34(4–5):532–58.
46. Sharma Y, Deepak K, Kumar P, Chauhan A. Blockage removal and RF controlled pipe inspection robot (BRICR). Int J Electr Electron Eng Telecomun. 2015;4(3):62–8.

47. Tadakuma K, Ming A, Shimojo M, Yoshida K, Keiji Nagatani Kazuya Yoshida Iagnemma K. Basic running test of the cylindrical tracked vehicle with sideways mobility. In: Intelligent robots and systems. IROS 2009. IEEE/RSJ international conference;2009, p. 1679–1684.
48. IPEK International GmbH. ROVVER Brochure. http://www.ipek.at/fileadmin/FILES/downloads/brochures/iPEK_rovver_web_en.pdf. Accessed 12 Oct 2013.
49. ABE Complex Technology. Kamery Inspekcyjne Samojezdne. http://abe-group.eu/kamery-inspekcyjne/modele-kamer-inspekcyjnych-samojezdnych/. Accessed 25 Jan 2015.
50. Autonomous Solutions Inc. Chaos High Mobility Robot. 2016. http://www.asirobots.com/products/chaos/. Accessed 26 Apr 2016.
51. RedZone. Solo Unmanned Inspection Robot. http://www.redzone.com/products/solo-robots/. Accessed 21 Oct 2012.
52. Responder Multi-Sensor Inspection for Large Diameter pipe. 2012. http://www.redzone.com/multi-sensor-inspection-minimizes-risk-of-pipe-failure-in-houston/. Accessed 22 Oct 2012.
53. CUES. Cues Ultra Shorty III. 2012. http://www.cuesinc.com/UltraShortyIII.html. Accessed 20 Nov 2012.
54. PSO AS. Inuktun Versatrax 300. http://en.pso.no/main-menu/products/inspection-equipment/crawler-systems/inuktun-versatrax-300.
55. Carnegie Mellon University. Carnegie Mellon's Snake Robots Learn To Turn By Following the Lead of Real Sidewinders. http://www.cmu.edu/news/stories/archives/2015/march/snake-robots-follow-sidewinders.html. Accessed 31 Mar 2016.
56. Hydropulsion. Hydropulsion Vertical Crawler. 2012. http://www.hydropulsion.com/robotic-crawler-systems/vertical-crawler/vertical_crawler.pdf. Accessed 30 Oct 2012.
57. Inuktun Services Ltd. Versatrax Vertical Crawler. 2015. http://www.inuktun.com/crawler-vehicles/versatrax-vertical-crawler.html. Accessed 18 Apr 2016.
58. Zhang Y, Yan G. In-pipe inspection robot with active pipe-diameter adaptability and automatic tractive force adjusting. Mech Mach Theory. 2007;42(12):1618–31.
59. NEOVISION s.r.o. Jetty, Cleaning and inspectional robot for air-induction ducting. http://www.neovision.cz/sols/jetty.html. Accessed 31 Mar 2016.
60. Nagaya K, Yoshino T, Katayama M, Murakami I, Ando Y. Wireless piping inspection vehicle using magnetic adsorption force. IEEE/ASME Trans Mechatron. 2012;17(3):472–9.
61. Kim D-W, Park C-H, Kim H-K, Kim S-B. Force adjustment of an active pipe inspection robot. In: 2009 Iccas-Sice. 2009; p. 3792–3797.
62. Lee D, Park J, Hyun D, Yook G, Yang H. Novel mechanisms and simple locomotion strategies for an in-pipe robot that can inspect various pipe types. Mechanism Mach Theor. 2012;56:52–68.
63. Nayak A, Pradhan S.K. Design of a new in-pipe inspection robot. In: Procedia engineering 2014 97:2081–2091.
64. Kuwada A, Tsujino K, Suzumori K, Ka T. Intelligent actuators realizing snake-like small robot for pipe inspection. In: IEEE international symposium on MicroNanoMechanical and human science; 2006. p. 1–6.

Chapter 3
Design of a Pipeline Inspection Mobile Robot With an Active Adaptation Mechanism

Pipe inspection robots are mechatronic devices that incorporate specialized mechanical structure, actuators, dedicated electronic systems and custom control architecture that provide functionalities for condition assessment of pipes. Design concept of a mechanical structure of a robot is the initial phase that leads to evaluation of other subsystems. Mechanical design of this robot is the continuation of a conceptual project, covered in master book, conducted by the author [1]. The project, described in the book is further developed and improved for manufacturing of a prototype.

The main objective of the design, presented in this doctoral book, was to create a robot with an active drive adaptation mechanism that would be able to transform for operation in wide variety of environments. It was motivated by the need to optimize pipeline inspection tasks and develop a versatile robot that can inspect places that would normally require several dedicated devices. The design process featured elaboration of an initial concept of mechanical model, followed by mechatronic design procedure of mechanical, electrical, electronic and control components of the robot.

3.1 Mechanical Structure

The mechanical structure of the robot was created, according to several assumptions: adaptable driving mechanism, operation in pipes with round and rectangular cross-section, motion in vertical pipes, driving on flat surfaces, suitability for water environment, integration with video inspection camera, remote operation by tether cable [1]. In order to fulfil the requirements, it was concluded, based on the market research that the most suitable driving mechanism for this application would be track drives. Moreover, to provide adaptation to different shapes of pipes, it is not possible to use motors located inside of the robot body to actuate the tracks, but it is necessary to utilize integrated track drive units. Additionally, to minimize robot size and

M. Ciszewski et al., *Modeling and Control of a Tracked Mobile Robot for Pipeline Inspection*, Mechanisms and Machine Science 82, https://doi.org/10.1007/978-3-030-42715-3_3

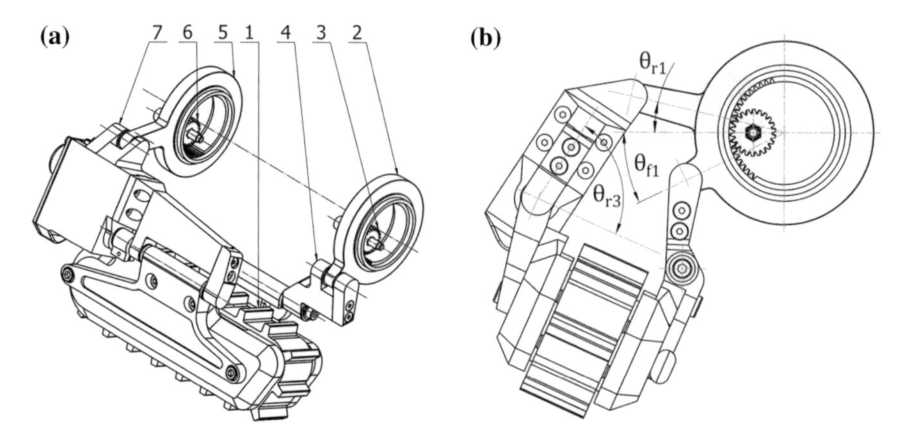

Fig. 3.1 Pedipulator mechanism [2]: 1–track drive module; 2–front driven ring; 5–rear driven ring; 3, 6–sprockets; 4–front arm; 7–rear driven arm; θ_r1–rear ring (5) rotation angle; θ_{f1}–front ring (2) rotation angle; θ_{r3}–rear arm (7) rotation angle with respect to the track drive module (1)

enhance manoeuvrability on flat surfaces or rough terrain, single-segment design was used, featuring rigid main body. The original design of the robot adaptable driving mechanism is based on two pedipulators that control the pose of track drive modules, as depicted in Fig. 3.1. Robot pedipulators can be interpreted as components of the mechanism that realize dexterous manipulations that resemble capabilities of human legs, in this case represented by track drive modules. Each pedipulator consists of two coaxial, independently driven actuated rings (2, 5), with an axis of rotation in the center of the robot body. These rings are driven by servomotors, located in the robot body from which torque is transferred by sprockets (3, 6) of internal meshing gear transmissions. Arms (4, 7) are mounted to the rings by revolute joints. Other sides of the arms are attached to the track drive module (1) by rotary joints. The rear arm (7) is equipped with an additional servomotor that sets orientation of the module (1) with respect to the arm.

By assembling two pedipulators to the robot body, a reconfigurable drive mechanism was constructed that allows positioning of track drives to provide robot motion in various environments [3]. The robot is equipped with six servomotors, responsible for setting proper pose of two track drive modules, thus the robot has 8 drives in total. Synbook of the mechanism was conducted iteratively, taking into account multiple requirements that appeared due to operation environment restrictions, available actuators sizes and technical parameters. The mechanism is scalable to larger pipe sizes by exchanging components such as rotating rings (2, 5) and arms (4, 7). The assumed dimensions were optimized for maximum versatility of the robot motion unit. Mechanical construction of the robot driving mechanism is protected by Polish patent [2].

Key criterion for material selection was mass optimization of the robot assembly. The robot body, front and rear arms, track drive mounts and camera supports were designed as aluminum components that feature low density, good machinability and sufficient corrosion resistance. The rotating rings were designed as stainless steel

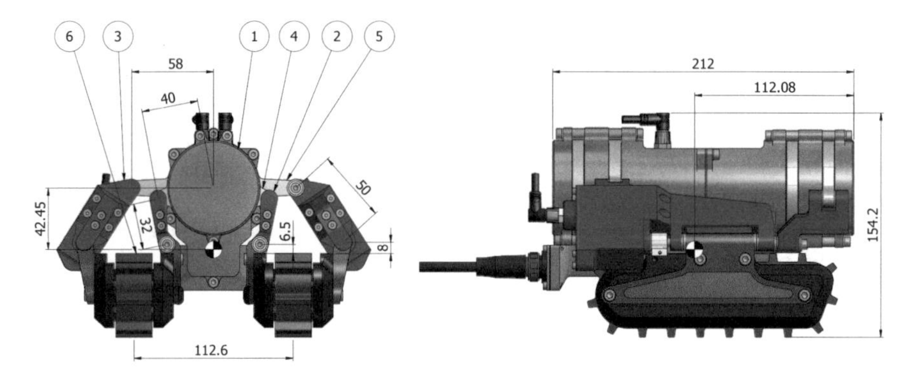

Fig. 3.2 Robot CAD model–general view: 1–robot body; 2–front arm; 3–rear arm; 4–front rotating ring; 5–rear rotating ring; 6–track drive module

parts due to higher strength requirements for integrated gear transmissions. Stainless steel screws and connectors were used in the robot to guarantee corrosion resistance.

The design of the robot ensures that it can operate in pipelines with active fluid flow, sewage pipes and other environments where watertightness and dust protection is required. Robot body compartments are protected by seals formed from aramide fiber gasket sheets. The rotating rings are sealed by rubber O-rings and cable routing is realized with usage of waterproof connectors rated at IP67 or IP68 ingress protection levels.

The CAD model of the robot is depicted in Fig. 3.2. In total, the robot assembly consists of over 230 components, among which over 60 have custom design and have to be manufactured by specialized workshop. Length of the robot body is 220 mm excluding inspection equipment and connectors. Height of the robot assembly differs, depending on the attained pose of the track drive modules.

3.2 Actuators

The track drive modules utilized for the robot are Inuktun Microtracs [4]. These modules that have IP68 rating are dedicated for usage in inspection robots and can be operated up to 30 m underwater. To decrease mass of the robot, aluminum version of the track drives was selected. Dimensions of the track drive unit are: $170 \times 50 \times 60$ mm. The tracks feature 65 N pull rating, 105 N payload capacity and maximum velocity of 0.15 m/s. They are actuated by brushed motors, supplied by 24 V DC voltage, rated at 75 W with planetary gear transmissions [4].

Poses of the track drives with respect to the robot body are set with usage of six servomotors. For actuation of the rotating rings, located in the robot body, Hitec HS-7950TH digital servomotors are utilized. They are equipped with coreless brushed DC motors and feature holding torque of 3.4 Nm and rotational velocity reaching 460 deg/s. Torque from each of the 4 servomotors is delivered individually to the rotating rings by internal meshing spur gear transmissions. Servomotors located in the rear robot arms that set orientation of these arms with respect to the track drives require higher holding torque due to the fact that they are utilized as direct drives. For this application, Hitec HS-7980TH model was selected that provides holding torque of 4.5 Nm and rotational velocity of 350 deg/s. Both servomotor models can be supplied with 6.0–7.4 V DC voltage. A negative feedback loop is implemented in the servomotors that utilizes position error information from analog potentiometers, fed to PID motor position controllers.

3.3 Inspection Equipment

The robot is equipped with an inspection camera that is mounted in the front of the body with usage of an adjustable mount. Its position can be changed from high to low, depending on the application and camera pitch angle can be adjusted. Due to size optimization, it was decided to use a constant focus camera with non-rotational head. The focus range is from 50 mm to infinity, whilst horizontal view angle is 100°. The head has resolution of 520 TVL that corresponds to 720 × 576 effective pixels. The camera is placed in an IP68 rated housing. It is additionally equipped with built-in 8-LED illumination that features externally adjustable brightness [5].

For operation in larger spaces, the robot is accessorized with an additional light, mounted in the front center of the body. It features a 10 W power LED with 60° wide beam lens, 930 lumens maximum luminous flux, waterproof housing and manually adjustable pitch angle.

3.4 Operation Environments

According to the design requirements, the robot is capable of positioning of its driving mechanism in various ways to accommodate to work environment. For the most compact alignment, it will be able to operate in pipes with diameter greater than Ø210 mm (Fig. 3.3a). In Fig. 3.3b, we may observe the robot in a pose for operation in a Ø350 mm diameter pipe. The upper limit of pipe diameter is determined by capabilities of the vision system. The robot may also work in pipes and ducts with rectangular cross-section. The minimum width of such pipe is 230 mm (Fig. 3.3c). As in the case of pipes with a circular cross-section, the maximum size is dependent on the capabilities of the utilized camera and lighting.

Visualization of the robot operation in horizontal pipes with pedipulators adaptation to diameters Ø210 and Ø315 mm is shown in Fig. 3.4.

Parallel extension of tracks is also possible for the robot structure. It may be utilized to operate in pipes or ducts with rectangular or circular cross-sections that are oriented in any direction, in particular vertically. Extension forces exerted by

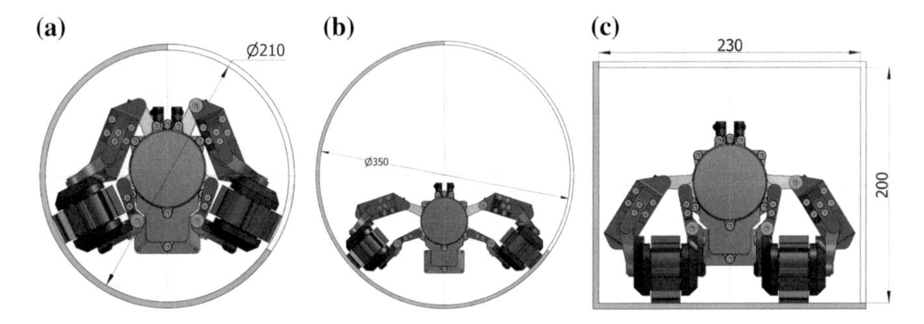

Fig. 3.3 Operation environments–front view: **a** Ø210 mm pipe; **b** Ø350 mm pipe; **c** rectangular duct 230 mm wide

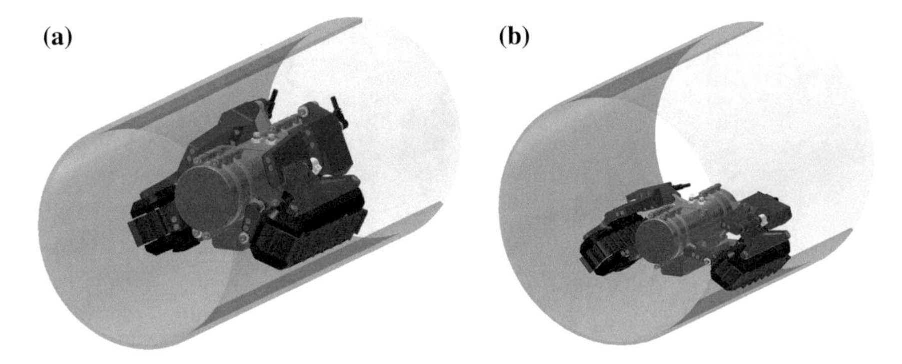

Fig. 3.4 Operation in horizontal pipes: **a** Ø210 mm; **b** Ø315 mm

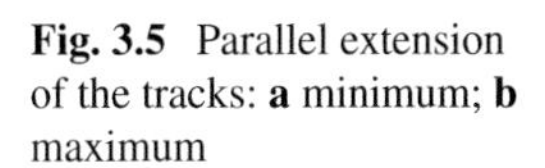

Fig. 3.5 Parallel extension of the tracks: **a** minimum; **b** maximum

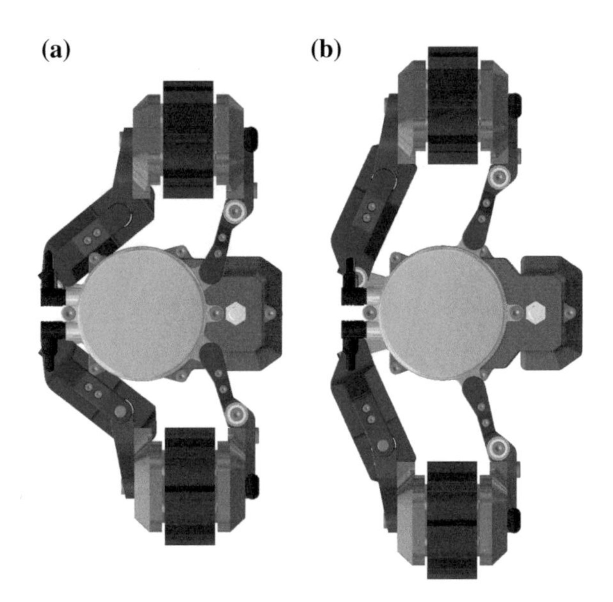

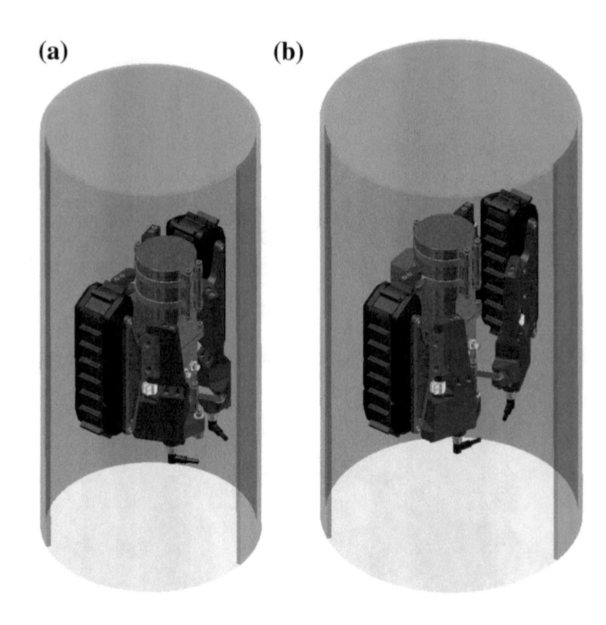

Fig. 3.6 Operation in vertical pipes: **a** Ø224 mm; **b** Ø270 mm

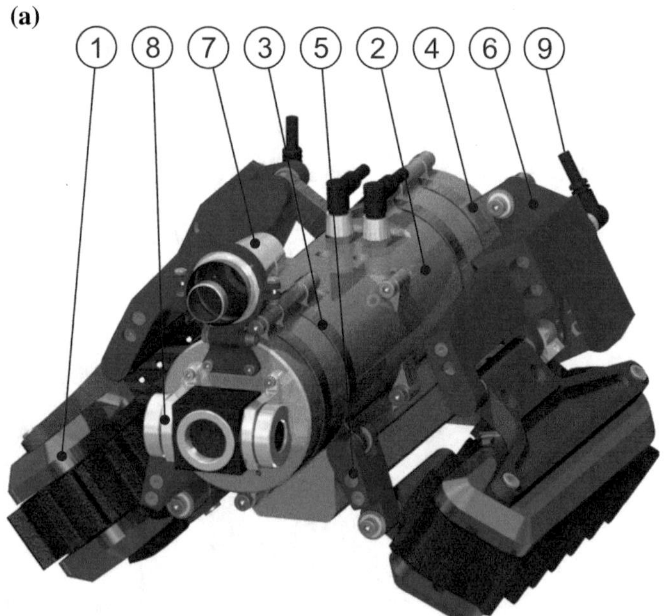

Fig. 3.7 Complete robot model: 1–track drive module; 2–robot body; 3–front rotating ring; 4–rear rotating ring; 5–front arm; 6–rear arm; 7–camera; 8–light; 9–waterproof connector

the pedipulators induce friction between rubber tracks and pipe walls, permitting the robot to propel forwards and backwards. Possible minimum and maximum extensions of the motion unit (from 224 to 270 mm) are presented in Figs. 3.5 and 3.6.

3.5 Complete Robot Model

A complete 3D CAD model of the robot is presented in Fig. 3.7. The robot is depicted in a pose for motion in horizontal pipe with circular cross-section. The CCTV camera is assembled in upper mounting position, on a dedicated support with adjustable tilt angle. The additional light is attached in the front of the robot body. It is also possible to attach the camera in lower mounting position, below the light to enhance Images/image quality of lower segments of inspected pipes.

The complete model of the robot was utilized as the documentation source for creation of working drawings, assembly drawings, stress analyzes, multibody simulations and component lists for manufacturing and building of a prototype.

3.6 Summary

Design of the pipe inspection robot with active adaptation mechanism proved that it is possible to construct a versatile mechatronic device that may be used for inspection of changing environments. Design objective was focused on the original pedipulator structure with closed kinematic chain. Mechanical construction of the robot was described, actuator specifications were given and video inspection equipment was outlined. On the basis of the 3D robot model, possible configurations and thus possible operation environments were visualized and commented. A complete, detailed 3D model of the robot was shown with all components that served as an indispensable resource for modeling and simulations that helped in control system design.

References

1. Ciszewski M. Tracked mobile robot for pipe and duct inspection. MA thesis, AGH University of Science and Technology; 2012.
2. Giergiel J, Giergiel M, Buratowski T, Ciszewski M. Mechanizm pedipulatora do ustawiania pozycji modułu napedowego, zwłaszcza robota mobilnego. PL2238752016.
3. Allegro Microsystems LLC. Current sensor ACS712. 2014. http://www.allegromicro.com/~/media/Files/Datasheets/ACS712-Datasheet.ashx?la=en.
4. Inuktun Services Ltd. Microtracs. 2014. http://www.inuktun.com/crawler-tracks/microtracs.html. Accessed 12 Sep 2014.
5. Allwan Security. Icam125 color camera. 2014. http://www.allwan.fr/Inspection/UW-3060.htm. Accessed 15 Oct 2012.

Chapter 4
Mathematical Modeling of the Robot

Mathematical modeling of the robot may be divided into three main sections. The first objective was to create a kinematic model of the robot motion on even surfaces. This model, described in [1] is an extension of kinematic modeling of skid-steering vehicles, since the robot's chassis is constantly arranged in a pose with parallel tracks. The model, however, does not provide information on the adaptation of the robot's motion unit to the environment, since motion is described on planar surfaces. Secondly, dynamic model of the robot motion on even surfaces is presented. This model is useful for control of the robot motion on even surfaces, including water environments and changing operating conditions. To complement the modeling approach, the most important model for the robot was developed that covers adaptation of the robot chassis to work environment. An original modeling approach led to creation of kinematic models of the pedipulators that set pose of the track drive modules. Forward and inverse kinematic equations of pedipulators closed kinematic chains are derived. An original algorithm for calculation of pedipulators transformation trajectories is developed that serves as a core functionality in simulations, development and implementation of the robot control system.

4.1 Kinematic Model of the Robot Motion on Even Surfaces

In this section, robot motion model for even surfaces is derived. The model is valid for constant pose of the robot's chassis with the track drive modules oriented parallelly. Construction of a kinematic model of the robot is commenced from analysis of track drive motion. Description of motion of a crawler track in a real environment with uneven ground and changeable conditions is very complex. Detailed mathematical description of movement of individual crawler track points is compound and it is advised to apply simplified models. Elastomer tracks with treads could be modeled as a non-stretch tape, wound about a determined shape by a drive sprocket, an idler

M. Ciszewski et al., *Modeling and Control of a Tracked Mobile Robot for Pipeline Inspection*, Mechanisms and Machine Science 82, https://doi.org/10.1007/978-3-030-42715-3_4

and an undeformable ground [2–4]. The presented kinematic model of the robot describes a plane motion and an operation on inclined surfaces.

The velocity of the point C (Fig. 4.1), placed in the axis of symmetry of the robot, [3, 4] may be expressed as:

$$V_c = \sqrt{\dot{x}_C^2 + \dot{y}_C^2 + \dot{z}_C^2} \tag{4.1}$$

where: V_c—velocity of the point C; $\dot{x}_C, \dot{y}_C, \dot{z}_C$—velocity components with respect to axes x, y, z.

The equations for particular velocity components were derived, taking into consideration different slip of the tracks and an assumption that the principal direction of motion is the x axis. Additionally, slope inclination by angle γ was assumed as a rotation of motion plane about the y axis and. The angle of turn β is positive from the x towards the y axis (Fig. 4.1):

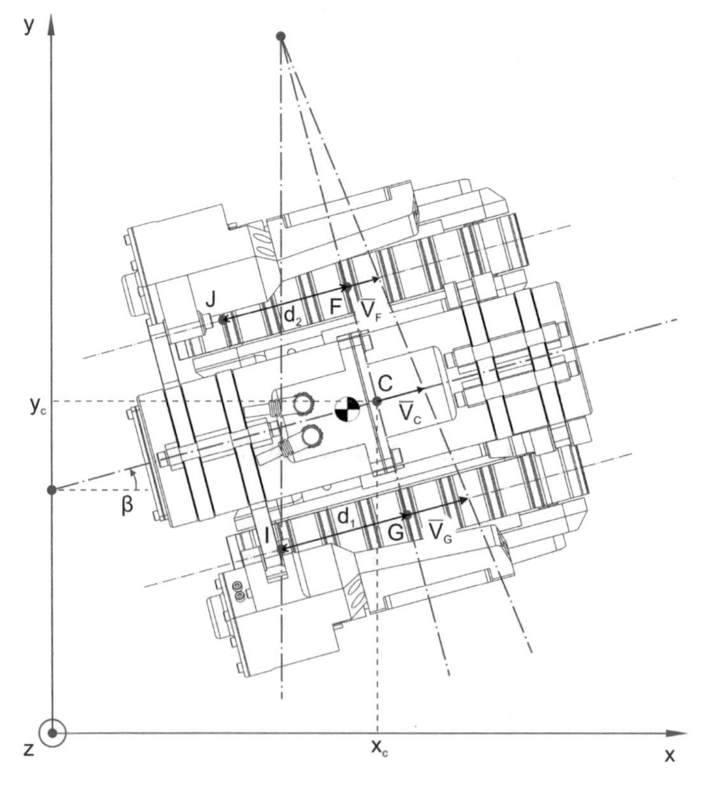

Fig. 4.1 Kinematic model of the robot motion on even surfaces

$$\begin{aligned} \dot{x}_C &= \tfrac{r\dot{\alpha}_1(1-s_1)+r\dot{\alpha}_2(1-s_2)}{2} cos\beta cos\gamma \\ \dot{y}_C &= \tfrac{r\dot{\alpha}_1(1-s_1)+r\dot{\alpha}_2(1-s_2)}{2} sin\beta \\ \dot{z}_C &= \tfrac{r\dot{\alpha}_1(1-s_1)+r\dot{\alpha}_2(1-s_2)}{2} sin\gamma \\ \dot{\beta} &= \tfrac{r\dot{\alpha}_1(1-s_1)-r\dot{\alpha}_2(1-s_2)}{H} \end{aligned} \tag{4.2}$$

where: r—radius of the track drive sprockets, H—distance between the tracks, s_1*—slip of the track 1,* s_2*—slip of the track 2, G—gravity force,* η*—efficiency,* $\dot{\alpha}_1$*—angular velocity of the sprocket 1,* $\dot{\alpha}_2$*—angular velocity of the sprocket 2,* γ*—angle of slope inclination.*

When a track load-bearing segment is in contact with the ground, then the effect of slip occurs. The slip phenomenon is affected by properties of the ground, driving force, type and placement of track treads. The driving force appearing in the robot track modules, exerts shear stresses on the ground. It is possible to determine the relationship between the driving force and factors that influence the slip [5]:

$$P_n = 10^6 b \int_0^L \tau_x dx. \tag{4.3}$$

where: P_n*—driving force; b—width of the crawler track; L—length of the load-bearing segment of the crawler track;* τ_x*—shear stresses in the soft ground defined by the Coulomb model*

By assuming that the course of parallel deformations of the ground is linear, it is possible to express these deformations by [6]:

$$\Delta l_x = x \cdot s \tag{4.4}$$

where: s—slip; x*—distance of the point for which the slip is calculated from the point of crawler track contact with the ground*

The greatest slip occurs for x = L, therefore, it is possible to express the slip by [6]:

$$s = \frac{\Delta l_x}{x} = \frac{\Delta l_{max}}{L} \tag{4.5}$$

The slip of a track is calculated using the following formula [6]:

$$s = \frac{(n-1) \cdot dL}{L} \tag{4.6}$$

where: n—number of track treads in contact with the ground; dL—track tread deformation; L—length of a track load bearing segment.

The velocities V_F and V_G of the points F and G, located in the centers of the tracks may be expressed as:

$$\begin{aligned} V_F^2 &= \dot{x}_F^2 + \dot{y}_F^2 + \dot{z}_F^2 \\ V_G^2 &= \dot{x}_G^2 + \dot{y}_G^2 + \dot{z}_G^2 \end{aligned} \tag{4.7}$$

$$\begin{cases} \dot{x}_F = \dot{x}_C - 0,5H\dot{\beta}sin\beta \\ \dot{y}_F = \dot{y}_C - 0,5H\dot{\beta}cos\beta \\ \dot{z}_F = \dot{z}_C \end{cases} \tag{4.8}$$

$$\begin{cases} \dot{x}_G = \dot{x}_C + 0,5H\dot{\beta}sin\beta \\ \dot{y}_G = \dot{y}_C + 0,5H\dot{\beta}cos\beta \\ \dot{z}_G = \dot{z}_C \end{cases} \tag{4.9}$$

The created kinematic model can be used to describe motion of a skid-steering tracked vehicle in a concise, computationally effective form. It will be used in determination of dynamic equations of robot motion on flat and inclined surfaces.

4.2 Dynamic Model of the Robot Motion on Even Surfaces

Dynamic description of the robot [2, 4, 7] was prepared using an energetic method based on the Lagrange equations, as presented in [8]. In order to avoid modeling problems with decoupling Lagrange multipliers, the Maggi's equations were used. In the dynamic model of the robot, described in [5, 8] the same characteristic points on the structure are considered as in the kinematic description (Fig. 4.2).

In the dynamic model, initially, kinetic energy is calculated and then all external forces are taken into consideration, thus potential energy is not explicitly specified. It has to be assumed that the kinetic energy of the robot E is the sum of energies of particular components, by application of the rule of kinetic energy summation [9]:

$$E = E_R + E_{M1} + E_{M2} \tag{4.10}$$

Fig. 4.2 Dynamic model of the robot—forces acting on the robot

where: E_R*—kinetic energy of the robot frame;* E_{M1}*—kinetic energy of the left track drive module;* E_{M2}*—kinetic energy of the right track drive module.*

Kinetic energy of the robot frame E_R is the sum of energies E_{R1} and E_{R2}, resultant from translational and rotational motions with respect to the instantaneous center of rotation O [9].

$$E_R = E_{R1} + E_{R2} = \frac{1}{2} m_R V_C^2 + \frac{1}{2} I_R \dot{\beta}^2 \tag{4.11}$$

where: m_R*—mass of the robot frame;* I_R*—moment of inertia of the robot frame;* $\dot{\beta}$*—angular velocity of the robot frame with respect to the instantaneous center of rotation.*

By introducing Eq. (4.1) into (4.11), the kinetic energy of the robot frame was obtained:

$$E_R = \frac{1}{2} m_R \left(\dot{x}_C^2 + \dot{y}_C^2 + \dot{z}_C^2\right) + \frac{1}{2} I_R \dot{\beta}^2 \tag{4.12}$$

Kinetic energy of the track drive module was determined with usage of the following formula:

$$E_M = E_{K1} + E_{K2} + E_{K3} + E_O \tag{4.13}$$

where: E_{K1}*—kinetic energy of the track drive sprocket 1;* E_{K2}*—kinetic energy of the idler 2;* E_{K3}*—kinetic energy of the idler 3;* E_O*—kinetic energy of the track module housing.*

Kinetic energy of the sprocket and idlers in the track module can be expressed as a sum of kinetic energies of translational motion, rotational motion about particular axis of rotation and rotational motion about the instantaneous center of rotation. The moments of inertia were determined for the particular models of the sprocket and idlers that were modeled in a CAD software, according to the datasheet from the Inuktun company [10].

$$\begin{aligned}
E_{K1} &= \frac{1}{2} m_{K1} V_A^2 + \frac{1}{2} I_{x1} \dot{\alpha}_{K1}^2 + \frac{1}{2} I_{z1} \dot{\beta}^2 \\
E_{K2} &= \frac{1}{2} m_{K2} V_B^2 + \frac{1}{2} I_{x2} \dot{\alpha}_{K2}^2 + \frac{1}{2} I_{z2} \dot{\beta}^2 \\
E_{K3} &= \frac{1}{2} m_{K3} V_E^2 + \frac{1}{2} I_{x3} \dot{\alpha}_{K3}^2 + \frac{1}{2} I_{z3} \dot{\beta}^2
\end{aligned} \tag{4.14}$$

where: m_{Ki}*—mass of i-th wheel;* $I_{x\,i}$*—moment of inertia with respect to i-th axis of rotation x;* $I_{z\,i}$*—moment of inertia of i-th wheel with respect to the axis z; about which the wheel changes its orientation with the angular velocity* $\dot{\beta}$*;* $\dot{\alpha}_{Ki}$*—angular velocity of i-th wheel;* V_A, V_B, V_E*—velocities of characteristic points presented in* Fig. 4.1.

The kinetic energy of the track module housing is the sum of energies of the motor, the gear transmission and the track.

$$E_O = \frac{1}{2}m_O V_O^2 + \frac{1}{2}I_{xO}\dot{\alpha}_1^2 + \frac{1}{2}I_{zO}\dot{\beta}^2 \tag{4.15}$$

where: m_O—mass of the track module housing; I_{xO}—moment of inertia of the elements in rotational motion; I_{zO}—moment of inertia of the housing with respect to the instantaneous center of rotation.

The total kinetic energy of one track drive module is denoted as follows:

$$\begin{aligned} E_M = {}& \tfrac{1}{2}m_{K1}V_A^2 + \tfrac{1}{2}I_{x1}\dot{\alpha}_{K1}^2 + \tfrac{1}{2}I_{z1}\dot{\beta}^2 + \tfrac{1}{2}m_{K2}V_B^2 + \tfrac{1}{2}I_{x2}\dot{\alpha}_{K2}^2 + \tfrac{1}{2}I_{z2}\dot{\beta}^2 + \\ & + \tfrac{1}{2}m_{K3}V_E^2 + \tfrac{1}{2}I_{x3}\dot{\alpha}_{K3}^2 + \tfrac{1}{2}I_{z3}\dot{\beta}^2 + \tfrac{1}{2}m_O V_O^2 + \tfrac{1}{2}I_{xO}\dot{\alpha}_1^2 + \tfrac{1}{2}I_{zO}\dot{\beta}^2 \end{aligned} \tag{4.16}$$

With assumption that velocities of characteristic points are equal:

$$V_A = V_B = V_E = V_O = V \tag{4.17}$$

$$\begin{aligned} E_M = {}& \tfrac{1}{2}V^2\left(m_{K1} + m_{K2} + m_{K3} + m_O\right) + \tfrac{1}{2}I_{x1}\dot{\alpha}_{K1}^2 + \tfrac{1}{2}I_{x2}\dot{\alpha}_{K2}^2 + \\ & + \tfrac{1}{2}I_{x3}\dot{\alpha}_{K3}^2 + \tfrac{1}{2}I_{xO}\dot{\alpha}_1^2 + \tfrac{1}{2}\dot{\beta}^2\left(I_{z1} + I_{z2} + I_{z3} + I_{zO}\right) \end{aligned} \tag{4.18}$$

When taking into account the relations between angular velocities and radii of the sprocket and idlers:

$$\begin{aligned} \alpha_{K1}r_1 &= \alpha_{K2}r_2 = \alpha_{K3}r_3 = \alpha_1 r \\ \dot{\alpha}_1 r_{K1} &= \dot{\alpha}_{K2}r_2 = \dot{\alpha}_{K3}r_3 = \dot{\alpha}_1 r \end{aligned} \tag{4.19}$$

Thus, with usage of the following substitution:

$$\begin{aligned} m &= m_{K1} + m_{K2} + m_{K3} + m_O \\ I_x &= I_{x1} + I_{x2}\left(\tfrac{r_1}{r_2}\right)^2 + I_{x3}\left(\tfrac{r_1}{r_3}\right)^2 + I_{xO} \\ I_z &= I_{z1} + I_{z2} + I_{z3} + I_{zO} \end{aligned} \tag{4.20}$$

The total kinetic energy for the track drive module was derived:

$$E_M = \frac{1}{2}mV^2 + \frac{1}{2}I_x\dot{\alpha}_1^2 + \frac{1}{2}I_z\dot{\beta}^2 \tag{4.21}$$

Previously, only one track drive module was investigated and particular properties were denoted without an index. However, in the more detailed analysis, the energy of the left and right track drive modules is utilized (according to the notation in Fig. 4.3):

$$\begin{aligned} E_{M1} &= \tfrac{1}{2}mV_F^2 + \tfrac{1}{2}I_x\dot{\alpha}_1^2 + \tfrac{1}{2}I_z\dot{\beta}^2 \\ E_{M2} &= \tfrac{1}{2}mV_G^2 + \tfrac{1}{2}I_x\dot{\alpha}_2^2 + \tfrac{1}{2}I_z\dot{\beta}^2 \end{aligned} \tag{4.22}$$

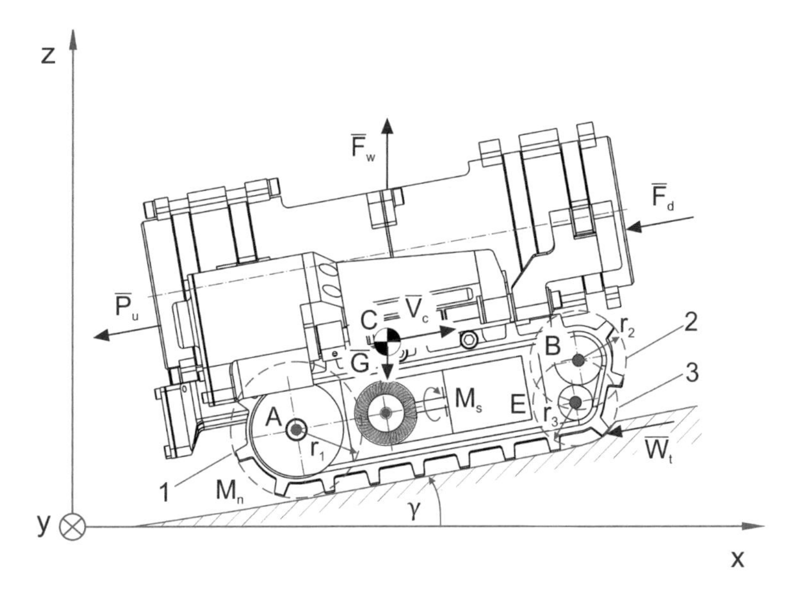

Fig. 4.3 Dynamic model of the robot—forces acting on the robot on an inclined surface *Where: 1—track drive sprocket; 2,3—idlers; r_1—radius of the sprocket; W_{t1}, W_{t2}—rolling friction forces; P_u—tether cable pull force; F_w—buoyant force; F_D—hydrostatic resistance force; M_n—torque on the drive sprocket; M_s—track motor torque; M_P—moment of transverse resistance; G—gravity force*

After substitution of velocities denoted in (4.7), the following formulas were obtained:

$$\begin{aligned} E_{M1} &= \tfrac{1}{2}m\left(\left(\dot{x}_C - \dot{\beta}Hsin\beta\right)^2 + \left(\dot{y}_C - \dot{\beta}Hcos\beta\right)^2 + \dot{z}_C^2\right) + \tfrac{1}{2}I_x\dot{\alpha}_1^2 + \tfrac{1}{2}I_z\dot{\beta}^2 \\ E_{M2} &= \tfrac{1}{2}m\left(\left(\dot{x}_C + \dot{\beta}Hsin\beta\right)^2 + \left(\dot{y}_C + \dot{\beta}Hcos\beta\right)^2 + \dot{z}_C^2\right) + \tfrac{1}{2}I_x\dot{\alpha}_2^2 + \tfrac{1}{2}I_z\dot{\beta}^2 \end{aligned} \tag{4.23}$$

The total kinetic energy of the robot described in (4.10) was derived with usage of Eqs. (4.12) and (4.23):

$$\begin{aligned} E = {} & \tfrac{1}{2}m_R\left(\dot{x}_C^2 + \dot{y}_C^2 + \dot{z}_C^2\right) + \tfrac{1}{2}I_R\dot{\beta}^2 + \tfrac{1}{2}m\left(\left(\dot{x}_c - \dot{\beta}Hsin\beta\right)^2 + \left(\dot{y}_C - \dot{\beta}Hcos\beta\right)^2 + \right. \\ & \left. + \ \dot{z}_C^2\right) + \tfrac{1}{2}I_x\dot{\alpha}_1^2 + I_z\dot{\beta}^2 + \tfrac{1}{2}m\left(\left(\dot{x}_C + \dot{\beta}Hsin\beta\right)^2 + \left(\dot{y}_C + \dot{\beta}Hcos\beta\right)^2 + \dot{z}_C^2\right) + \tfrac{1}{2}I_x\dot{\alpha}_2^2 \end{aligned} \tag{4.24}$$

In order to determine dynamic equations of motion, the Maggi's formalism was utilized [3, 11]:

$$\begin{aligned} \sum_{j=1}^{n} C_{ij}\left[\frac{d}{dt}\left(\frac{\partial E}{\partial \dot{q}_j}\right) - \left(\frac{\partial E}{\partial q_j}\right)\right] &= \Theta_i \\ \dot{q}_j &= \sum_{i=1}^{s} C_{ij}\dot{e}_i + G_j \end{aligned} \tag{4.25}$$

where: $i = 1, \ldots, s$—number of independent parameters, expressed in generalized coordinates q_j, in the number equal to degrees of freedom of the system; $j = 1, \ldots, n$—number of generalized coordinates; E—kinetic energy; C_{ij}—coefficient matrix; G_j—coefficients; Θ_i—generalized forces; $\dot{e}_i$—kinetic parameters in generalized coordinates

With usage of the principle of virtual work of kinetic parameters $\delta\dot{e}_i$, we can express Eq. (4.25) in external forces Q_j [3]:

$$\sum_{i=1}^{s} \delta\dot{e} C_{ij} Q_{j_i} = \sum_{i=1}^{s} \delta\dot{e} \sum_{i=1}^{s} C_{ij} Q_j \tag{4.26}$$

With usage of the Maggi's formalism, in analogous way to a 2-wheeled mobile robot, the kinetic parameters $\dot{e}_i$ and coefficients G_j were obtained as follows [11]:

$$\begin{aligned} \dot{e}_i &= [\dot{\alpha_1}\ \dot{\alpha_2}]^T \\ G_j &= [0\ 0\ 0\ 0\ 0\ 0]^T \end{aligned} \tag{4.27}$$

According to this assumption, by Eq. (4.26), six generalized velocities $\dot{q}_j$ were denoted by multiplication of the matrix C_{ij}, that consists of nonholonomic constraints with two kinetic parameters $\dot{\alpha}_1, \dot{\alpha}_2$.

$$\begin{bmatrix} \dot{x}_C \\ \dot{y}_C \\ \dot{z}_C \\ \dot{\beta} \\ \dot{\alpha}_1 \\ \dot{\alpha}_2 \end{bmatrix} = \begin{bmatrix} \frac{1}{2}r\,(1-s_1)\,sin\beta & \frac{1}{2}r\,(1-s_2)\,sin\beta \\ \frac{1}{2}r\,(1-s_1)\,cos\beta cos\gamma & \frac{1}{2}r\,(1-s_2)\,cos\beta cos\gamma \\ \frac{1}{2}r\,(1-s_1)\,sin\gamma & \frac{1}{2}r\,(1-s_2)\,sin\gamma \\ -\frac{r(1-s_1)}{H} & \frac{r(1-s_2)}{H} \\ 1 & 0 \\ 0 & 1 \end{bmatrix} \begin{bmatrix} \dot{\alpha}_1 \\ \dot{\alpha}_2 \end{bmatrix} \tag{4.28}$$

The generalized forces and moments Θ_i are denoted by Eq. (4.29), when equilibrium equations of forces and moments are derived [9], according to notation given in Figs. 4.2 and 4.3:

$$\Theta_i = \begin{bmatrix} M_{n1} + (-0,5P_u - 0,5F_D - 0,5Gsin\gamma + 0,5F_w sin\gamma - 0,5W_{t1})\,r\,(1-s_1) + M_p\frac{r(1-s_1)}{H} \\ M_{n2} + (-0,5P_u - 0,5F_D - 0,5Gsin\gamma + 0,5F_w sin\gamma - 0,5W_{t2})\,r\,(1-s_2) - M_p\frac{r(1-s_2)}{H} \end{bmatrix} \tag{4.29}$$

By composition of Eqs. (4.24), (4.25), (4.28) and (4.29), the final form of the dynamic motion equations, based on the Maggi's formalism, have been presented as follows [8]:

$$\begin{pmatrix} \frac{r}{2}\left[\ddot{\alpha}_1(1-s_1)+\ddot{\alpha}_2(1-s_2)\right]sin\beta + \\ +\frac{r}{2}\left[\dot{\alpha}_1(1-s_1)+\dot{\alpha}_2(1-s_2)\right]\frac{r\dot{\alpha}_2(1-s_2)-r\dot{\alpha}_1(1-s_1)}{H}cos\beta \end{pmatrix}(m_R+2m)\,\frac{1}{2}r\,(1-s_1)\,sin\beta +$$
$$+\begin{pmatrix} \frac{r}{2}\left[\ddot{\alpha}_1\,(1-s_1)+\ddot{\alpha}_2\,(1-s_2)\right]cos\beta cos\gamma - \\ +\frac{r}{2}\left[\dot{\alpha}_1\,(1-s_1)+\dot{\alpha}_2\,(1-s_2)\right]\frac{r\dot{\alpha}_2(1-s_2)-r\dot{\alpha}_1(1-s_1)}{H}sin\beta cos\gamma \end{pmatrix}(m_R+2m)\,\frac{1}{2}r\,(1-s_1)\,\cdot$$
$$\cdot\, cos\beta cos\gamma + \left(\frac{r}{2}\left[\ddot{\alpha}_1(1-s_1)+\ddot{\alpha}_2(1-s_2)\right]sin\gamma\right)(m_R+2m)\,\frac{1}{2}r\,(1-s_1)\,sin\gamma -$$
$$+\left(\frac{r\ddot{\alpha}_2(1-s_2)-r\ddot{\alpha}_1(1-s_1)}{H}\right)\left(I_R+2I_z+2mH^2\right)\frac{r(1-s_1)}{H}+I_x\ddot{\alpha}_1$$
$$= M_{n1} + (-0,5P_u - 0,5F_D - 0,5Gsin\gamma + 0,5F_w sin\gamma - 0,5W_{t1})\,r\,(1-s_1) + M_p\frac{r(1-s_1)}{H} \quad (4.30)$$

$$\begin{pmatrix} \frac{r}{2}\left[\ddot{\alpha}_1(1-s_1)+\ddot{\alpha}_2(1-s_2)\right]sin\beta + \\ +\frac{r}{2}\left[\dot{\alpha}_1(1-s_1)+\dot{\alpha}_2(1-s_2)\right]\frac{r\dot{\alpha}_2(1-s_2)-r\dot{\alpha}_1(1-s_1)}{H}cos\beta \end{pmatrix}(m_R+2m)\,\frac{1}{2}r\,(1-s_2)\,sin\beta +$$
$$+\begin{pmatrix} \frac{r}{2}\left[\ddot{\alpha}_1(1-s_1)+\ddot{\alpha}_2(1-s_2)\right]cos\beta cos\gamma - \\ +\frac{r}{2}\left[\dot{\alpha}_1(1-s_1)+\dot{\alpha}_2(1-s_2)\right]\frac{r\dot{\alpha}_2(1-s_2)-r\dot{\alpha}_1(1-s_1)}{H}sin\beta cos\gamma \end{pmatrix}(m_R+2m)\,\frac{1}{2}r\,(1-s_2)\,\cdot$$
$$\cdot cos\beta cos\gamma + \left(\frac{r}{2}\left[\ddot{\alpha}_1(1-s_1)+\ddot{\alpha}_2(1-s_2)\right]sin\gamma\right)(m_R+2m)\,\frac{1}{2}r\,(1-s_2)\,sin\gamma +$$
$$+\left(\frac{r\ddot{\alpha}_2(1-s_2)-r\ddot{\alpha}_1(1-s_1)}{H}\right)\left(I_R+2I_z+2mH^2\right)\frac{r(1-s_2)}{H}+I_x\ddot{\alpha}_2$$
$$= M_{n2} + (-0,5P_u - 0,5F_D - 0,5Gsin\gamma + 0,5F_w sin\gamma - 0,5W_{t2})\,r\,(1-s_2) - M_p\frac{r(1-s_2)}{H} \quad (4.31)$$

where: r—radius of the track drive sprocket; α_1, α_2—*angles of rotations of the sprockets 1 and 2*; m_R—*mass of the frame*; *m—mass of the track*; W_t—*rolling friction force*; P_u—*tether cable pull force*; F_w—*buoyant force*; F_D—*hydrostatic resistance force*; M_{n1}, M_{n2}—*torque on the drive sprockets of the tracks 1,2; H—distance between the tracks*; I_R—*moment of inertia of the robot frame*; I_X, I_Z—*reduced moments of inertia of the track drive module*; M_P—*moment of transverse resistance;* s_1, s_2—*slip of tracks 1 and 2*; *G—gravity force*; η—*efficiency*; γ—*slope inclination*

The dynamic equations of motion (4.30) and (4.31) may be used to solve forward and inverse dynamics problems for the tracked mobile robot. However, care must be taken when calculating the values of forces, particularly the rolling friction force W_t as various surfaces on which the robot operates would introduce significant variations in its value. Type of fluid in which the robot moves has also strong influence on the forces, especially on F_D and M_P. Slip of the tracks that is dependent on surface condition also plays important role in these calculations.

4.3 Kinematic Model of the Pedipulators

Previous sections were devoted to description of kinematic and dynamic models of the robot motion on even surfaces. In case of pipelines, where environment restricted by pipe walls is generally encountered during motion, application of complex models to perform normal inspection tasks is not crucial. However, the original design of the robot's chassis, based on two pedipulators requires custom modeling and control approach to transform to different work environments.

Mathematical model of the pedipulators is essential for proper operation of the robot, because transformation trajectories are used to adapt the robot's chassis to different sizes of pipes, even surfaces and rough terrain. Forward and inverse kinematics for pedipulators closed kinematic chains are presented in this section.

4.3.1 Forward Kinematics

In general, the pedipulator structure for adaptation of one track module is a closed kinematic chain that consists of six rotational joints of 5th class and five links. It should be noted that in this case 5th class of the kinematic pairs for the rotational joints means that five degrees of freedom are taken, therefore only one motion (rotation in this case) is possible, according to nomenclature given by [12]. The mechanism can be treated as planar, due to the fact that the rotational joints axes are parallel. The axis of rotation of coaxial rings that move the robot arms was assumed as the z axis of the coordinate system and the axes x and y are oriented according to the right-hand rule, as shown in Fig. 4.4.

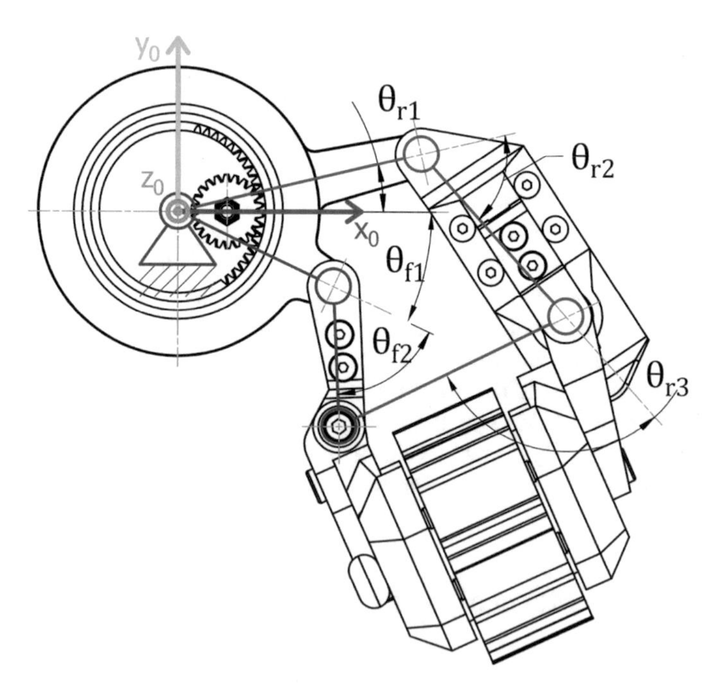

Fig. 4.4 Kinematic scheme of the left pedipulator. Division of the kinematic chain into a rear 3-DOF manipulator (red) and a front 2-DOF manipulator (blue)

The mobility of a mechanism in space can be expressed using Eq. (4.32) [12].

$$w = 6n - \sum_{i=1}^{i=5} ip_i \tag{4.32}$$

where: n—number of links; i—*class of the kinematic pair*; p_i—*number of kinematic pairs of i-th class*

When the mechanism of one pedipulator is analyzed in terms of mobility w, according to Eq. (4.32), we obtain the theoretical mobility given by (4.33).

$$w = 6 \cdot n - 5 \cdot p_5 = 6 \cdot 5 - 5 \cdot 6 = 0 \tag{4.33}$$

The theoretical mobility calculated in this method does not correspond to the real mobility, therefore, since the mechanism is planar, the equation for mobility of planar mechanisms, given by Eq. (4.34) can be used [12].

$$w = 3n - 1 \cdot p_4 - 2 \cdot p_5 = 3 \cdot 5 - 0 - 2 \cdot 6 = 3 \tag{4.34}$$

In this case, the mobility that is equal to 3 correspond to the real mobility of one pedipulator. In this mechanism, rotational joints, providing change of angles θ_{r1} and θ_{f1}, corresponding to rotation of rings are actuated by servomotors, located in the robot body. The third servomotor that adjusts angle θ_{r3} is located in the rear arm of the pedipulator. Other joints are non-actuated.

In order to apply a modeling approach, widely used in robotics for open kinematic chains, the mechanism was divided into two planar manipulators with two and three degrees of freedom that represent front and rear parts of the pedipulator, shown in Fig. 4.4. Additionally, it was assumed that the manipulators have to move their end-effectors concurrently to retain positions in the same point that would fulfil constraints of the closed kinematic chain. If the mechanism was treated as a closed kinematic chain with five links, the analyses would be much more complicated due to increasing amount of geometric redundancies. The pedipulator was also analyzed as an open kinematic chain with five links but this method proved to be less stable and computationally inefficient. Generally, it is computationally more difficult to analyze closed kinematic chains [12] and for this pedipulator mechanism, usage of the alternative approach was justified. Thus, for both planar manipulators that were formed after division, transformation matrices were created using standard Denavit-Hartenberg notation [13]. A homogeneous transformation matrix that is formed to describe position and orientation is denoted by (4.35). Its short form is represented by Eq. (4.36) [13].

$$T_{n,0}(q) = \begin{bmatrix} r_{11} & r_{12} & r_{13} & p_x \\ r_{21} & r_{22} & r_{23} & p_y \\ r_{31} & r_{32} & r_{33} & p_z \\ 0 & 0 & 0 & 1 \end{bmatrix} \tag{4.35}$$

where: $T_{n,0}$*—homogeneous transformation matrix from n-th coordinate system to base coordinate system;* q*—vector of generalized joint variables;* $r_{11} \ldots r_{33}$*—orientation of n-th coordinate system with respect to base coordinate system;* p_x, p_y, p_z*—position*

$$T_{n,0}(q) = \begin{bmatrix} R_{n,0}(q) & p_{n,0}(q) \\ 0 & 1 \end{bmatrix} \tag{4.36}$$

where: $R_{n,0}(q) - 3 \times 3$ *rotation matrix;* $p_{n,0}(q)$*—*3×1 *position vector.*

Table 4.1 presents Denavit-Hartenberg (DH) parameters for the front 2-DOF manipulator and Table 4.2 shows the parameters for the rear manipulator.

The transformation from end-effector to the base coordinate system for the front (2-DOF) manipulator is presented by Eq. (4.37) and for the rear (3-DOF) manipulator by Eq. (4.38). At this stage of calculations, it was assumed that all rotational joints are actuated. It is not valid for the real mechanism, but without this assumption it would be impossible to use this modeling approach. Later, for control purposes, real mechanism structure would be recovered in the model and additional joints, assumed as actuated in the kinematic modeling would be freed. This operation would be possible with application of supplementary conditions and geometric constraints.

$$T^F_{2,0} = \begin{bmatrix} cos(\theta_{f1}+\theta_{f2}) & -sin(\theta_{f1}+\theta_{f2}) & 0 & a_{f2}cos(\theta_{f1}+\theta_{f2}) + cos(\theta_{f1})a_{f1} \\ sin(\theta_{f1}+\theta_{f2}) & cos(\theta_{f1}+\theta_{f2}) & 0 & a_{f2}sin(\theta_{f1}+\theta_{f2}) + sin(\theta_{f1})a_{f1} \\ 0 & 0 & 1 & 0 \\ 0 & 0 & 0 & 1 \end{bmatrix} \tag{4.37}$$

$$T^R_{3,0} = \begin{bmatrix} cos(\theta_{r1}+\theta_{r2}+\theta_{r3}) & -sin(\theta_{r1}+\theta_{r2}+\theta_{r3}) & 0 & a_{r3}cos(\theta_{r1}+\theta_{r2}+\theta_{r3}) + a_{r2}cos(\theta_{r1}+\theta_{r2}) + cos(\theta_{r1})a_{r1} \\ sin(\theta_{r1}+\theta_{r2}+\theta_{r3}) & cos(\theta_{r1}+\theta_{r2}+\theta_{r3}) & 0 & a_{r3}sin(\theta_{r1}+\theta_{r2}+\theta_{r3}) + a_{r2}sin(\theta_{r1}+\theta_{r2}) + sin(\theta_{r1})a_{r1} \\ 0 & 0 & 1 & 0 \\ 0 & 0 & 0 & 1 \end{bmatrix} \tag{4.38}$$

Table 4.1 Denavit-Hartenberg parameters for the 2-DOF manipulator

No.	θ_i	$\mathbf{d_i}$	$\mathbf{a_i}$	α_i
1	θ_{f1}^{var}	0	a_{f1}	0
2	θ_{f2}^{var}	0	a_{f2}	0

where: θ_{f1}, θ_{f2}*—rotation of joints 1, 2;* a_{f1}, a_{f2}*—lengths of links 1, 2*

Table 4.2 Denavit-Hartenberg parameters for the 3-DOF manipulator

No.	θ_i	$\mathbf{d_i}$	$\mathbf{a_i}$	α_i
1	θ_{r1}^{var}	0	a_{r1}	0
2	θ_{r2}^{var}	0	a_{r2}	0
3	θ_{r3}^{var}	0	a_{r3}	0

where: $\theta_{r1}, \theta_{r2}, \theta_{r3}$*—rotation of joints 1, 2, 3;* a_{r1}, a_{r2}, a_{r3}*—lengths of links 1, 2, 3*

For the entire motion, position constraints between end-effectors of both manipulators must be maintained. The constraint is denoted by Eq. (4.39) and after substitution by (4.40). It ensures closure of the open kinematic chains to form a proper closed kinematic chain that represents the real structure. It should be noted that the usage of forward kinematics equations is limited due to the loop closure constraints. It is necessary to use a specific approach for control of the pedipulator position and orientation that involves iterative inverse kinematics calculations with constraints denoted by (4.40)

$$\begin{cases} T^F_{2,0}\,(1,4) = T^R_{3,0}\,(1,4) \\ \\ T^F_{2,0}\,(2,4) = T^R_{3,0}\,(2,4) \end{cases} \tag{4.39}$$

$$\begin{cases} a_{f2}cos\left(\theta_{f1}+\theta_{f2}\right) + a_{f1}cos\left(\theta_{f1}\right) - (a_{r3}cos\,(\theta_{r1}+\theta_{r2}+\theta_{r3}) + \\ \qquad + a_{r2}cos\,(\theta_{r1}+\theta_{r2}) + a_{r1}cos\,(\theta_{r1})\,) = 0 \\ \\ a_{f2}sin\left(\theta_{f1}+\theta_{f2}\right) + a_{f1}sin\left(\theta_{f1}\right) - (a_{r3}sin\,(\theta_{r1}+\theta_{r2}+\theta_{r3}) + \\ \qquad + a_{r2}sin\,(\theta_{r1}+\theta_{r2}) + a_{r1}sin\,(\theta_{r1})\,) = 0 \end{cases} \tag{4.40}$$

4.3.2 Inverse Kinematics

Inverse kinematics task is the most important calculation procedure for control of the pedipulators. Classical analytical methods cannot be simply used for control of the pedipulators due to complex mechanical structure, existence of redundancies and kinematic constraints that arise in this closed kinematic chain. Thus, a dedicated, original calculation algorithm was developed to realize this task. Initially, analytical and geometric calculations for a particular pedipulator pose will be described, next numerical differential kinematics with usage of manipulator Jacobian will be shown and finally, an effective calculation procedure will be described that utilizes all methods to generate and validate trajectories for pose control of the pedipulators.

The analytical approach to solution of inverse kinematics problems for 2-DOF and 3-DOF planar manipulators are well-known procedures. However they provide several solutions for the same position of the end-effectors [13]. In order to solve inverse kinematics problem for the pedipulators, it was necessary to use analytical solution for the 3-DOF manipulator as a supplementary source of conditions, utilized in the calculation procedure. A scheme of the 3-DOF manipulator, utilized for this task is shown in Fig. 4.5.

Forward kinematics equations for this manipulator are denoted by (4.38). The transformation matrix gives position of end-effector with respect to the base coordinate system: $p_{3x} = \mathrm{T}^{\mathrm{R}}_{3,0}\,(1,4)$ and $p_{3y} = \mathrm{T}^{\mathrm{R}}_{3,0}\,(2,4)$. The orientation is dependent on the sum of joint angles and we can denote the angle as (4.41).

$$\theta_r = \theta_{r1} + \theta_{r2} + \theta_{r3} \tag{4.41}$$

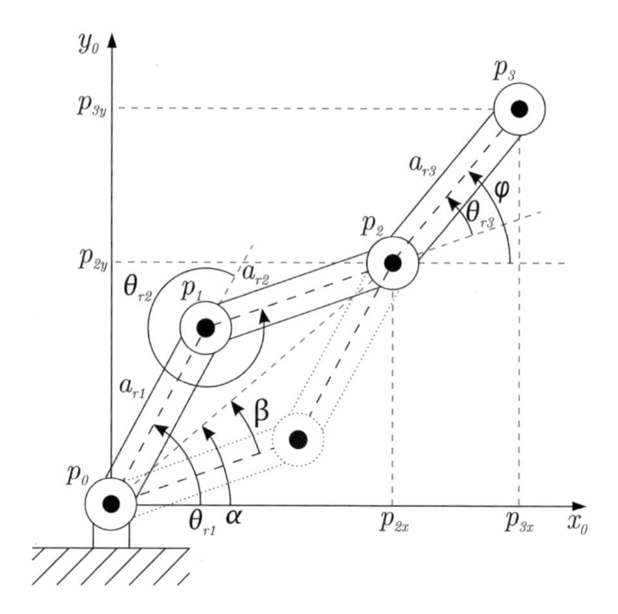

Fig. 4.5 Scheme of the 3-DOF planar manipulator: geometric solution to the inverse kinematics problem

If we do not specify the angle θ_r, we can get infinite number of solutions for position given by p_{3x} and p_{3y}. First, let's consider the position at the end of second link denoted by p_{2x} and p_{2y} (4.42).

$$\begin{cases} p_{2x} = p_{3x} - a_{r3}cos\,(\theta_r) & = a_{r1}cos\,(\theta_{r1}) \ + a_{r2}cos?(\theta_{r1} + \theta_{r2}) \\ p_{2y} = p_{3y} - a_{r3}sin\,(\theta_r) & = a_{r1}sin\,(\theta_{r1}) \ + a_{r2}sin?(\theta_{r1} + \theta_{r2}) \end{cases} \tag{4.42}$$

If we square and sum both sides of Eq. (4.42), we can obtain:

$$p_{2x}^2 + p_{2y}^2 = a_{r1}^2 + a_{r2}^2 + 2a_{r1}a_{r2}cos\,(\theta_{r2}) \tag{4.43}$$

With usage of (4.43) we can determine functions cos (θ_{r2}) and sin?(θ_{r2}) and use them to calculate the second joint rotation:

$$\theta_{r2} = Atan2\,(sin\,(\theta_{r2}), cos\,(\theta_{r2})\,) \tag{4.44}$$

When we substitute solution of (4.44) to (4.42), we can obtain solution for the first rotational joint angle θ_{r1}:

$$\theta_{r1} = Atan2\,(sin\,(\theta_{r1}), cos\,(\theta_{r1})\,) \tag{4.45}$$

This value can be found by usage of the Foglowie functions:

$$\begin{aligned} cos\left(\theta_{r1}\right) &= \frac{\left(a_{r1}+a_{r2}cos\left(\theta_{r2}\right)\right)p_{2y}-a_{r2}p_{2x}cos\left(\theta_{r2}\right)}{p_{2x}^2+p_{2y}^2} \\ sin\left(\theta_{r1}\right) &= \frac{\left(a_{r1}+a_{r2}cos\left(\theta_{r2}\right)\right)p_{2x}-a_{r2}p_{2y}cos\left(\theta_{r2}\right)}{p_{2x}^2+p_{2y}^2} \end{aligned} \tag{4.46}$$

The third rotational joint angle θ_{3r} can be found with usage of substitution to Eq. (4.47):

$$\theta_{r3}=\left(\theta_r-\theta_{r1}-\theta_{r2}\right) \tag{4.47}$$

This operation finishes analytical solution to the inverse kinematics problem. However, even when we determine the angle θ_r, there exist two solutions representing different orientations of first two arms. For control of the pedipulator, it is required to select particular solution, valid for the mechanical structure. Therefore, geometric solution to the problem was derived.

Additional angles α and β were introduced in order to represent angles in the triangles that will be analyzed. First, the Law of Cosines can be used to the triangle formed by manipulator arms with lengths a_{r1} and a_{r2} and a section OP_1, assuming that Eq. (4.48) is fulfilled. It is true when the position lays within reach of the second arm [13].

$$\sqrt{p_{2x}^2+p_{2y}^2}\leq a_{r1}+a_{r2}. \tag{4.48}$$

$$p_{2x}^2+p_{2y}^2=a_{r1}^2+a_{r2}^2+2a_{r1}a_{r2}cos\left(\pi-\theta_{r2}\right) \tag{4.49}$$

When we solve for cos?(θ_{r2}) we get the same result as in previous method:

$$cos\left(\theta_{r2}\right)=\frac{p_{2x}^2+p_{2y}^2-a_{r1}-a_{r2}}{2a_{r1}a_{r2}} \tag{4.50}$$

Depending on the sign of the angle θ_{r2}, we can have two possible solutions. For $\theta_{r2}\in(-\pi,0)$, we will have the manipulator arm in the upward position, whereas for $\theta_{r2}\in(0,\pi)$, we will have the arm in downward position. To find the variable θ_{r1}, we can use the angles α represented by (4.51) and β that can be derived with usage of Law of Cosines (4.52) and transformed to final representation (4.53).

$$\alpha=Atan2(p_{2x},p_{2y}) \tag{4.51}$$

$$cos\left(\beta\right)\sqrt{p_{2x}^2+p_{2y}^2}=a_{r1}+a_{r2}cos\left(\theta_{r2}\right) \tag{4.52}$$

$$\beta=arccos\left(\frac{p_{2x}^2+p_{2y}^2+a_{r1}^2-a_{r2}^2}{2a_{r1}\sqrt{p_{2x}^2+p_{2y}^2}}\right) \tag{4.53}$$

The value of $\beta \in (0,\pi)$ was derived in order to make particular triangles. The angle θ_{r1} can be calculated by the Eq. (4.54).

$$\begin{cases} \theta_{r1} = \alpha + \beta, & \theta_{r2} > 0 \\ \theta_{r1} = \alpha - \beta, & \theta_{r2} < 0 \end{cases} \tag{4.54}$$

Finally, we can calculate the third angle with usage of Eq. (4.47). In order to validate calculations of the inverse kinematics, various pedipulator poses were checked numerically using forward and inverse kinematics models derived analytically.

Velocity relationships between position and orientation of links with respect to the base coordinate system are denoted by derivatives of the transformation matrix. A notion of manipulator Jacobian is used in differential kinematics. A Jacobian is the matrix equivalent of the derivative—the derivative of a vector-valued function of a vector with respect to a vector [14]. For a manipulator with n degrees of freedom, a Jacobian matrix with dimensions $6 \times n$ is used. The first three rows of the Jacobian represent transformation of linear velocity, whereas, the last three contain information about angular velocity. We can denote the spatial velocity V using (4.55) [14].

$$V = \left[v_x, v_y, v_z, \omega_x, \omega_y, \omega_z\right]^T \in \mathbb{R}^6 \tag{4.55}$$

where: v_x, v_y, v_z*—linear velocities*; $\omega_x, \omega_y, \omega_z$*—angular velocities*

To derive a manipulator Jacobian, notion of a skew-symmetric matrix for time-varying rotation matrix has to be used. Derivative of a time-varying rotation matrix, that represents the rate of change of orientation in time can be represented as (4.56), where for a 3-dimensional case, the skew-symmetric matrix is denoted by (4.57).

$$\dot{R}(t) = S(\omega)R(t) \tag{4.56}$$

where: $\dot{R}(t)$*—*3×3 *time-varying rotation matrix*; $S(\omega)$*—skew-symmetric matrix*

$$S(\omega) = \begin{bmatrix} 0 & -\omega_z & \omega_y \\ \omega_z & 0 & -\omega_x \\ -\omega_y & \omega_x & 0 \end{bmatrix} \tag{4.57}$$

Velocity with respect to the base coordinate system can be expressed by (4.58). In the equation, the geometric manipulator Jacobian J is used.

$$V = \begin{bmatrix} v_{n,0} \\ \omega_{n,0} \end{bmatrix} = \begin{bmatrix} J_v \\ J_\omega \end{bmatrix} \dot{q} = J\dot{q} \tag{4.58}$$

where: $v_{n,0}, \omega_{n,0}$*—linear and angular velocities*; J_v, J_ω*—Jacobian elements representing linear and angular velocities*; $\dot{q}$*—velocity of generalized joint coordinates (linear or angular)*

In order to calculate linear velocity components of manipulator Jacobian J_{υ}, we can use superposition rule to express velocity components that are dependent on all generalized joint coordinates using Eq. (4.59) [15].

$$p_{n,0} = \sum_{i=1}^{n} \frac{\partial p_{n,0}}{\partial q_i} \dot{q}_i \tag{4.59}$$

In case when a joint i is prismatic, the velocity of end-effector that is affected by the joint has the same magnitude and sense as the motion axis z_i. When a joint i is rotational and moves with angular velocity ω_i, the resultant linear velocity of the end-effector can be calculated using vector product of ω_i and the vector length between both coordinate systems $\upsilon_n = \omega_i \times p_{n,i}$. The column i in Jacobian J_{υ} will have the form described by Eq. (4.60). The value of vector $p_{i,0}$ is express by the last column of homogeneous transformation matrix $T_{i,0}$, whereas value z_i can be found as the third column of homogeneous transformation matrix [15].

$$\begin{cases} J_{\upsilon i} = R_{i,0}k = z_i & \textit{prismatic joint} \\ J_{\omega i} = z_i \times \left(p_{n,0} - p_{i,0}\right) & \textit{rotational joint} \end{cases} \tag{4.60}$$

In order to determine particular elements of manipulator Jacobian related to change of orientation, Eq. (4.56) has to be transformed and a general formula can be obtained for the J_{ω}, that can simplify calculation of the Jacobian [15]. An equation of angular velocity composition can be defined as (4.61).

$$\omega_{n,0} = \omega_{1,0} + R_{1,0}\omega_{2,1} + R_{2,0}\omega_{3,2} + \cdots + R_{n-1,0}\omega_{n,n-1} \tag{4.61}$$

When we take into consideration that a joint can be either rotational or prismatic, according to the Denavit-Hartenberg notation, we can write (4.62) and by substitution to Eq. (4.61), we can denote general equation of angular velocity as (4.63).

$$\begin{cases} \omega_{i,i-1} = 0 & \textit{prismatic joint} \\ \omega_{i,i-1} = \dot{q}_i k & \textit{rotational joint} \end{cases} \tag{4.62}$$

$$\omega_{n,0} = \sigma_1 \dot{q}_1 R_{1,0}k + \sigma_2 \dot{q}_2 R_{2,0}k + \cdots + \sigma_n \dot{q}_n R_{n,0}k = \sum_{i=1}^{n} \sigma_i \dot{q}_i z_i \tag{4.63}$$

where: σ_i—coefficient of joint type (0 for prismatic joint, 1 for rotational joint); k—versor of z_0 axis; $z_i = R_{i,0}k$—versor of axis of rotation i with respect to base coordinate system

With usage of Eq. (4.63), we can define manipulator Jacobian components of angular velocity as (4.64). It is only necessary to use versors of particular rotation axes defined by 3rd column of rotation matrix $R_{i,0}$.

$$J_{\omega} = (\sigma_1 z_1 \ldots \sigma_n z_n) \tag{4.64}$$

It should be noted that to determine geometric manipulator Jacobian matrix, it is sufficient to use the third and fourth columns of homogeneous transformation matrices $T_{i,0}$ and use Eqs. (4.60) and (4.64). The generalized method may be used to algorithmically calculate geometric manipulator Jacobian [16].

To solve inverse kinematics problem for the manipulators using differential kinematics, manipulator Jacobians were derived. Next, a method with Jacobian pseudoinverse, comparable to the ones described in [17, 18] was used for generation of joint trajectories between particular positions of tracks. Jacobian for the front manipulator is given by Eq. (4.65), whereas manipulator Jacobian for the rear manipulator is given by Eq. (4.66).

$$J^F = \begin{bmatrix} -a_{f2}sin\left(\theta_{f1}+\theta_{f2}\right) - sin\left(\theta_{f1}\right)a_{f1} & -a_{f2}sin\left(\theta_{f1}+\theta_{f2}\right) \\ a_{f2}cos\left(\theta_{f1}+\theta_{f2}\right) + cos\left(\theta_{f1}\right)a_{f1} & a_{f2}cos\left(\theta_{f1}+\theta_{f2}\right) \\ 0 & 0 \\ 0 & 0 \\ 0 & 0 \\ 1 & 1 \end{bmatrix} \tag{4.65}$$

$$J^R = \begin{bmatrix} -a_{r3}sin\left(\theta_{r1}+\theta_{r2}+\theta_{r3}\right) - a_{r2}sin\left(\theta_{r1}+\theta_{r2}\right) - sin\left(\theta_{r1}\right)a_{r1} & -a_{r3}sin\left(\theta_{r1}+\theta_{r2}+\theta_{r3}\right) - a_{r2}sin\left(\theta_{r1}+\theta_{r2}\right) & -a_{r3}sin\left(\theta_{r1}+\theta_{r2}+\theta_{r3}\right) \\ a_{r3}cos\left(\theta_{r1}+\theta_{r2}+\theta_{r3}\right) + a_{r2}cos\left(\theta_{r1}+\theta_{r2}\right) + cos\left(\theta_{r1}\right)a_{r1} & a_{r3}cos\left(\theta_{r1}+\theta_{r2}+\theta_{r3}\right) + a_{r2}cos\left(\theta_{r1}+\theta_{r2}\right) & a_{r3}cos\left(\theta_{r1}+\theta_{r2}+\theta_{r3}\right) \\ 0 & 0 & 0 \\ 0 & 0 & 0 \\ 0 & 0 & 0 \\ 1 & 1 & 1 \end{bmatrix} \tag{4.66}$$

When a manipulator Jacobian is a square and non-singular matrix, it is possible to find joint velocities with usage of Jacobian inversion (4.67). This method is valid only for 6-DOF manipulators in 3D space and is called resolved-rate motion control [14].

$$\dot{q} = J^{-1}\begin{bmatrix} v_{n,0} \\ \omega_{n,0} \end{bmatrix} \tag{4.67}$$

In case when a robot has number of DOF $n < 6$, we refer to it as under-actuated and when $n > 6$, it is called over-actuated or redundant. In case of an under-actuated robot, the Jacobian has lower number of columns than rows, thus we cannot use matrix inverse. A pseudo-inverse method can be used for under-actuated robots and has analogous property as matrix inverse (4.68). We can define the pseudo-inverse using (4.69). Therefore, for an under-actuated robot, the least squares solution, that minimizes computation error (4.70) for joint velocities is denoted using (4.71) [14].

$$J^{+}J = I \tag{4.68}$$

where: J^{+}—pseudo-inverse of Jacobian J matrix; *I—identity matrix*

$$J^{+} = \left(J^{T}J\right)^{-1}J^{T} \tag{4.69}$$

$$e = |\, J\dot{q} - V| \tag{4.70}$$

$$\dot{q} = J(q)^{+}V \tag{4.71}$$

where: $\dot{q}$—joint velocities; *$J(q)^{+}$—pseudo-inverse of Jacobian*; *V—velocity vector*

Another method that is suitable e.g. for 3-DOF planar manipulator is to remove Jacobian rows to decrease row number to the number of available degrees of freedom for particular structure, since other DOFs are not controllable [14]. This method utilizes a mask matrix of dimensions 1×6 that consists of elements equal to 1 or 0. It specifies Cartesian DOF (in the wrist coordinate frame) that will be ignored during calculation of a solution: $x,\ y,\ z,\ \omega_x,\ \omega_y, \omega_z$. For this application, mask matrix for rear 3-DOF manipulator denoted by Eq. (4.68) was used that allows to control x, y, ω_z coordinates of the end-effector.

$$M_R = [1\ 1\ 0\ 0\ 0\ 1] \tag{4.72}$$

Calculation of analytical representation of manipulator pseudo-inverse gives compound results, therefore numerical calculations of Jacobian for particular joint positions are used in practice. A general, numerical approach for calculation of inverse kinematics was used. It utilizes the Virtual Work principle, a vector of forces and moments and manipulator Jacobian to iteratively minimize error between initial and goal poses of the end-effector. With use of the principle, it is possible to state that the virtual work of all applied forces and moments is zero for all virtual movements of the system from static equilibrium [19] as denoted by Eq. (4.72).

$$\delta W = (F_E \cdot \delta p + M_E \cdot \delta\phi) - Q \cdot \delta q = G_E \cdot \delta\xi - Q \cdot \delta q = 0 \tag{4.73}$$

where: δW—virtual work, F_E—external end-effector forces, δp—virtual linear displacements of end-effector, M_E—external end-effector moments, $\delta\phi$—virtual angular displacements of end-effector, Q—generalized joint forces, δq—generalized virtual joint displacement, G_E—generalized external end-effector forces, $\delta\xi$—generalized virtual end-effector displacement (pose change)

The manipulator Jacobian can be used to transform joint velocity to an end-effector spatial velocity V, according to Eq. (4.58) and analogically, Jacobian transpose can be applied to transform generalized forces applied at the end-effector to torques and forces applied at the joints [14]. This method, whilst valid for dynamic modeling is useful in general inverse kinematics. It is based on minimization of error between initial and goal poses. The calculation procedure, described in [14] was accustomed for solution of the numerical inverse kinematics problem of the pedipulators. Generally, a pose can be denoted by Cartesian coordinates (4.74).

$$\xi = \left[p_x,\ p_y,\ p_z,\ \phi_x,\ \phi_y, \phi_z\right]^T \tag{4.74}$$

Generalized external forces applied to end-effector are expressed by (4.75).

$$G_E = [f_x,\ f_y,\ f_z,\ m_x,\ m_y, m_z] \tag{4.75}$$

With use of the principle of virtual work, and by Eq. (4.58), we can write:

$$G_E \cdot \left[\delta p_x\ \delta p_y\ \delta p_z\ \delta ?_x\ \delta ?_y\ \delta ?_z\right]^T = Q \cdot \delta q \tag{4.76}$$

$$\left[\delta p_x\ \delta p_y\ \delta p_z\ \delta \phi_x\ \delta \phi_y\ \delta \phi_z\right]^T = J(q)\delta q \tag{4.77}$$

The mapping of external forces to joint forces is realized with usage of manipulator Jacobian, thus for all virtual displacements, by combining Eqs. (4.76) and (4.77), we get:

$$G_E^T J\ (q)\ \delta q = Q^T \delta q \tag{4.78}$$

This equation must be valid for all virtual generalized displacements δq, therefore we obtain:

$$G_E^T J\ (q) = Q^T \tag{4.79}$$

Finally, external forces can be mapped to joint forces and torques with usage of Eq. (4.80). The mapping is never singular as it can be in the case of velocity kinematics using Jacobian approach. It is used for numerical solution of inverse kinematics problems [14].

$$Q = J^T G_E \tag{4.80}$$

In Fig. 4.6, a scheme of numerical inverse kinematics solution for the 3-DOF manipulator is shown. Actual manipulator pose is represented with solid outline, whereas desired pose is represented with dashed outline. The figure shows planar case of the problem. Analogous approach is used for solution of three-dimensional problems.

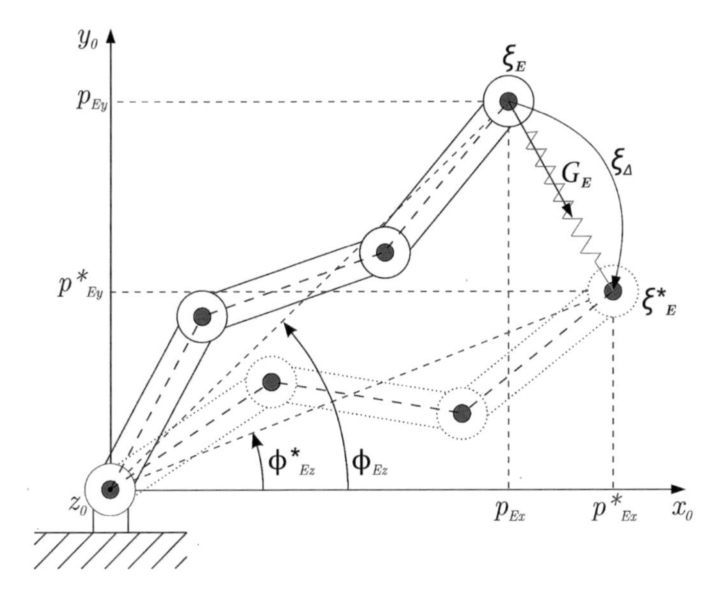

Fig. 4.6 Scheme of the 3-DOF manipulator inverse kinematics—numerical solution

where: $\xi_E(p_{Ex}, p_{Ey}, p_{Ez})$*—end-effector current pose;* ξ_E^* $(p_{Ex}^*, p_{Ey}^*, p_{Ez}^*)$*—end-effector desired pose;* ξ_Δ*—end-effector pose difference*

The principle of this method relies on a so called "special spring" between current and desired pose. This spring is a set of generalized external forces G_E that are able to change position and orientation of the end-effector towards the desired pose. It is proportional to the difference between poses (4.81), where proportionality is denoted by a constant γ_P. Current pose at a given calculation step $\xi_E \langle k \rangle$, is computed with usage of forward kinematics (4.82).

$$G_E \propto \Delta\left(\xi_E,\ \xi_E^*\right) \rightarrow G_E = \gamma_P \cdot \Delta\left(\xi_E,\ \xi_E^*\right) \tag{4.81}$$

$$\xi_E \langle k \rangle = T_{E,0}\left(q \langle k \rangle\right) \tag{4.82}$$

where: $q \langle k \rangle$*—current estimate of inverse kinematics solution,* $T_{E,0}$*—transformation matrix from end-effector to base coordinate system*

The end-effector external forces are resolved to joints forces with usage of Jacobian transpose, expressed in Eq. (4.80) and thus, the following formula is valid:

$$Q \langle k \rangle = J\left(q \langle k \rangle\right)^T \cdot G_E \langle k \rangle \tag{4.83}$$

By assumption that the virtual robot does not have joint motors, but only viscous dampers that impose proportionality of applied forces to joint velocity, we can write:

$$\dot{q}\langle k\rangle = Q\langle k\rangle / B \tag{4.84}$$

where: B—joint damping coefficients (the same for all joints)

Therefore, discrete time-update for the joint coordinates is given by:

$$q\langle k+1\rangle = \alpha_C \cdot \dot{q}\langle k\rangle + q\langle k\rangle \tag{4.85}$$

where: α_C—gain selected according to desired calculation convergence speed

The calculations are performed iteratively until the magnitudes of generalized external forces G_E that actuate the manipulator to desired pose are sufficiently small. With usage of this method, numerical determination of inverse kinematics can be executed. Depending on problem complexity, Jacobian pseudo-inverse can be used instead of transposition operation. In some cases, it can be more efficient than Jacobian transpose [20]. In problems when analyzed robot has less than 6-DOF, a mask vector, as in Eq. (4.72) can be applied, that gives:

$$Q\langle k\rangle = J\left(q\langle k\rangle\right)^T \cdot diag(M) \cdot G_E\langle k\rangle \tag{4.86}$$

where: diag(M)—function used to create a matrix with the elements of mask vector M on the main diagonal

Finally, after substitution of Eqs. (4.84), (4.85) and (4.86), discrete update of joint coordinates for an under-actuated robot is given by Eq. (4.87).

$$q\langle k+1\rangle = \alpha_C \cdot \frac{J\left(q\langle k\rangle\right)^T \cdot diag\left(M\right) \cdot G_E\langle k\rangle}{B} + q\langle k\rangle \tag{4.87}$$

The method specified above is used for numerical inverse kinematics calculations of the 3-DOF manipulator, subdivided from the robot pedipulator structure. It is integrated as a part of the calculation procedure, described in the trajectory calculation algorithm and plays important role in control system of the robot.

4.4 Trajectory Calculation Algorithm for Pedipulators Transformation

The generation of trajectory between arbitrarily selected poses of pedipulators is a complex task in case of this robot's motion unit structure. It was shown that usage of standard inverse kinematics approaches that would be applied to the entire kinematics chain may not lead to satisfactory results. Thus, an original calculation algorithm was developed that is composed of different methods utilized iteratively, numerical inverse kinematics based on of Jacobian pseudo-inverse, analytical inverse kinematics, forward kinematics, sets of geometric conditions and custom rules.

Pedipulator poses dedicated for particular pipe sizes were determined using the 3D CAD model of the robot. With usage of this data source, all pedipulators joints angular positions were saved and used as initial and final poses for trajectory generation. The second step was to generate 5-th order polynomial interpolated joint space trajectories for the front manipulator between initial and final poses. The 5-th order polynomial was selected since it gives smooth trajectory and minimizes jerk, whilst computational expenses are not excessive for this specific task. Higher order polynomials produce smoother trajectories, but require more mathematical operations [21]. Next, inverse kinematics problem was solved with usage of Jacobian pseudo-inverse for the rear manipulator to match its position with end-effector of the front manipulator. Similar approach for solution of the inverse kinematics problem of 3-DOF manipulator was presented by paper [22]. In case of this pipe inspection robot, calculation of the pedipulator analytical inverse kinematics problem gives 8 solutions for the entire structure. Therefore, additional conditions were used for the planar manipulators in an analogous manner to calculation procedure specified in [13]. Joint rotation ranges were also limited, according to mechanical structure. The final step was to eliminate oscillations of the manipulator, caused by transition through singular positions. Finally, joint trajectories of both 2-DOF and 3-DOF, artificially extracted planar manipulators were merged and only actuated joints were selected to perform reconfiguration of the pedipulators. With usage of simulations and validation tools, it was proved that appropriate trajectory generation with usage of two artificially extracted manipulators leads to evaluation of proper joint space trajectories for control of the pedipulator closed kinematic chain. The calculations were performed in MATLAB software with addition of the Robotics Toolbox [14]. As a result, smooth trajectories for transformation of both robot pedipulators were obtained to realize required motion in the workspace. Due to the fact that each pedipulator mechanism is a closed kinematics chain of specific structure, it is sufficient to use three motors to set particular pose of one track drive module [23].

The trajectory calculation algorithm is divided into two phases. The first, preparation phase includes operations that need to be executed only once to create database and a mathematical robot model. It consists of the following steps:

1. Setup the 3D robot model manually to adapt driving mechanism for motion in pipe of particular shape and size (preparation step executed only once for each required pose).
2. Read joint angular positions for a set of initial and final poses of the pedipulators.
3. Store joints angular positions in memory for all required pedipulator poses.
4. Split each pedipulator into 2-DOF (front) and 3-DOF (rear) planar manipulators.
5. Build kinematic models of the manipulators structure using DH-notation.

The second, execution phase includes operations that have to be processed for each trajectory generation task:

1. Select initial and final pedipulator poses (from predefined poses or transformation matrices).
2. Generate joint space trajectory between initial and final pose for both planar manipulators with usage of a quintic (5th order) polynomial with default zero boundary conditions for velocity and acceleration. In this operation, time vector

with defined number of steps is used [24]. Length of the vector is set, depending on required resolution of calculations.

3. Calculate temporary transformation matrices for each step of the generated trajectory using forward kinematics for front and rear manipulators (T_R^{tmp}, T_F^{tmp}).
4. Create temporary transformation matrix with orientation from rear manipulator and position from the front manipulator, thus satisfying condition of end-effector position of both manipulators given by (4.39) and (4.40). Maintain orientation of end-effector that complies to track drive module roll angle, specified by last link of the 3-DOF manipulator. The elements of temporary transformation matrices for each trajectory step are denoted by Eq. (4.88):

$$\begin{cases} T_R^{tmp2}\,(1:3, 1:3) = T_R^{tmp}\,(1:3, 1:3) \\ T_R^{tmp2}\,(1:3, 4) = T_F^{tmp}\,(1:3, 4) \end{cases} \tag{4.88}$$

5. Calculate rear manipulator joint positions using inverse kinematics for all poses defined by the temporary transformation matrices.
6. The calculations are performed with usage of numerical inverse kinematics methods that utilize Jacobian pseudo-inverse [25]. For consecutive trajectory steps, the inverse kinematics calculation is applied, where joint positions input for the next step is the solution from the previous step. The numerical inverse kinematics method is based on three-dimensional space. Thus, for underactuated robots that possess fewer than 6 DOF, the solution space has more dimensions than manipulator joint coordinates. In this case a mask vector (1×6) is used which specifies the Cartesian DOF (in the end-effector coordinate frame) that will be ignored in calculations of a solution. The mask vector has six elements that correspond to translation in X, Y and Z, and rotation about X, Y and Z axes respectively. In this particular procedure, the mask vector was set to consider X, Y positions and rotation with respect to Z axis. Equation (4.87) is used in this step.
7. Calculate modulus after division of joint angular position values to remove those that exceed 360°.
8. Calculate analytical solution for inverse kinematics for all transformation matrices of rear 3-DOF pedipulator, according to the procedure given by Eqs. (4.41)–(4.54).
9. Apply joint angular position correction with usage of analytical solution and joint rotation limits. First, the angular positions of angles θ_{r1} and θ_{r2} are checked with respect to the limits. Next, Eq. (4.41) is used to correct the angles. After that, the condition according to Eq. (4.54) is utilized and Eq. (4.47) is applied. Finally, the correction is also based on iterative checks of consecutive joint positions to eliminate sudden manipulator jerks caused by rapid angular position changes. An experimentally defined threshold is applied.
10. Store final joint trajectories for front 2-DOF and rear 3-DOF manipulators:

$$\begin{array}{c} q_f\,(:, 1:2) \\ q_r(:, 1:3) \end{array} \tag{4.89}$$

11. Select trajectories of actuated joints to control the pedipulator transformation. The actuated joints are: first joint of front manipulator, first and third joint of rear manipulator, according to Eq. (4.90). Only three drives are used to control position and orientation of the track modules, owing to the original structure of this closed kinematic chain. They are arranged as described in [23].

$$\begin{cases} q_p(:,1) = q_f(:,1) \\ q_p(:,2) = q_r(:,1) \\ q_p(:,3) = q_r(:,3) \end{cases} \tag{4.90}$$

12. Apply joint offsets for control the pedipulator motion from neutral pose and apply transmission ratio of internal meshing gear transmissions located in the robot body for front ring drive and rear ring drives:

$$\begin{cases} q_p(:,1) = (q_f(:,1) - q_f^{neu}(1)) \cdot i_{gear} \\ q_p(:,2) = (q_r(:,1) - q_r^{neu}(1)) \cdot i_{gear} \\ q_p(:,3) = (q_r(:,3) - q_r^{neu}(3)) \end{cases} \tag{4.91}$$

where: q_p, q_f, q_r*—pedipulator drives; front manipulator drives; rear manipulator drives angular positions—respectively;* i_{gear}*—internal meshing gears transmission ratio.*

Application of the algorithm presented above provides a valid solution for transformation of robot pedipulators in task space. It is possible to use poses predefined in CAD software as the input, but properly defined homogeneous transformation matrices can also be employed as input for the algorithm to allow e.g. track extension in vertical pipes, where track drive orientation about Z axis remains constant and position with respect to X axis is changed (Fig. 4.4). Other kinds of track drive pose interpolations are also possible, but explicit solution range may be limited due to intrinsic mechanism structure that imposes geometric constraints.

A flowchart that depicts workflow described by the algorithm is presented in Fig. 4.7. The "Joint trajectory calculation" is the most computationally demanding section in the procedure. Its execution time is strongly dependent on the desired initial and final poses. Successful operation of "Joint angular position correction using analytical solution" is also dependent on input data supplied to the algorithm. It should be noted that minimum duration of the trajectory is limited by maximum velocity of servomotors located in the pedipulators. However, for operation of the prototype it is advisable to increase duration of the transformation process to optimize power consumption and decrease jerks of the mechanism that can destabilize video recording and potentially reduce inspection footage quality.

On the basis of the previously presented mathematical models, implemented in the original algorithm, joint trajectories for different transformations were obtained. Transformation time was assumed to be 5 seconds for all cases to ensure easier comparison of the results. Positioning drives angles, angular velocities and accelerations were plotted for exemplary transformation trajectories, calculated with usage of the

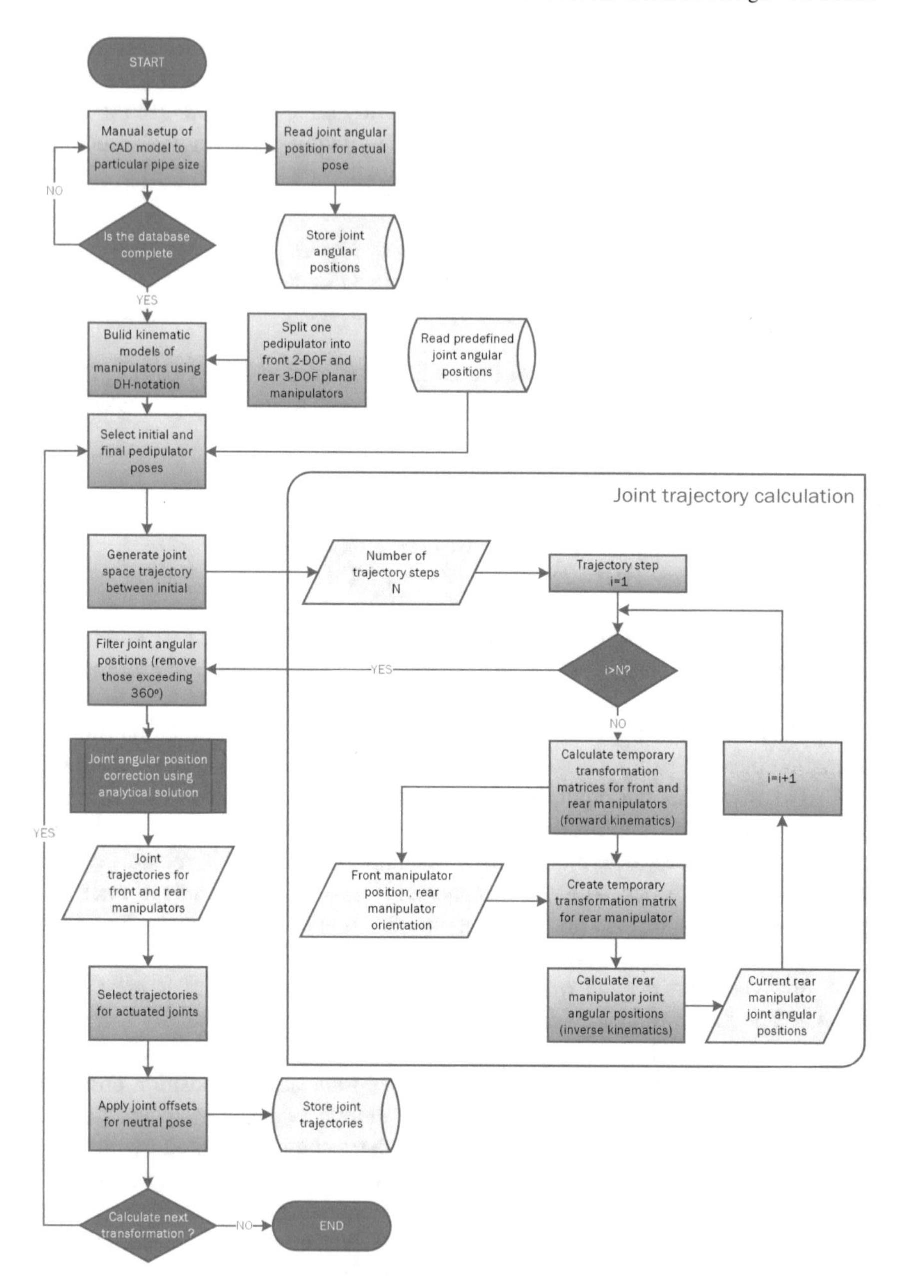

Fig. 4.7 Flowchart of the trajectory calculation algorithm for the pedipulators transformation

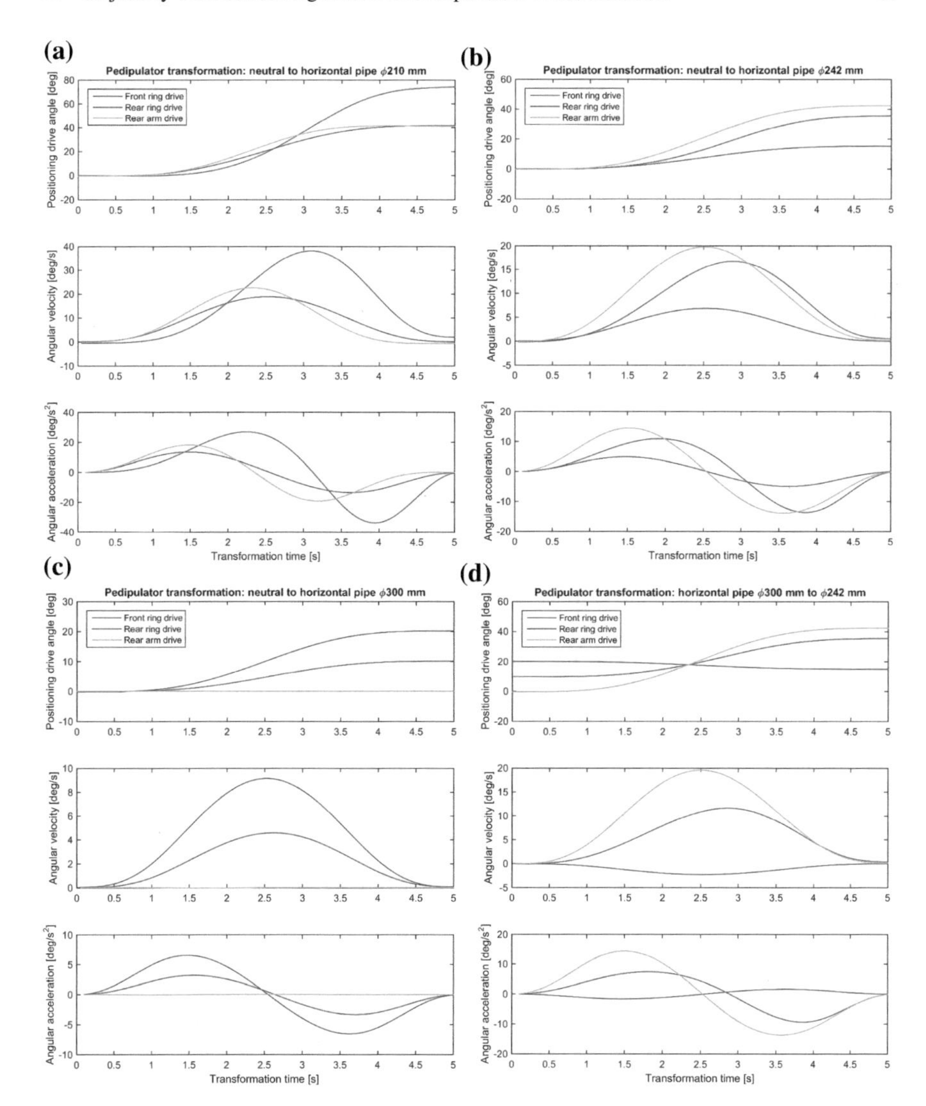

Fig. 4.8 Pedipulator transformations for horizontal pipes—drives positions, velocities and accelerations: **a** from the neutral pose to a ∅210 mm pipe; **b** from the neutral pose to a ∅242 mm pipe; **c** from the neutral pose to a ∅300 mm pipe; **d** from a ∅300 to a ∅242 mm pipe

algorithm. Calculation results of the pedipulator transformations for operation in horizontal pipes of selected internal diameters are depicted in Fig. 4.8.

The results for transformation from neutral poses to horizontal pipes with diameters ∅210 mm, ∅242 mm and ∅300 mm are depicted in Fig. 4.8a–c respectively. Figure 4.8d shows the results for transformation that the robot should perform during motion in pipe connector (reducer) that provides increase of internal diameter from

∅242 to ∅300 mm. The transformations depicted in Fig. 4.8a–c are used for initial setting of the robot before working in the pipes. For proper execution in simulation or on a prototype, they require a dedicated stand or a support of the body for unobstructed motion to realize pedipulator trajectories.

The use of 5-th order polynomial interpolation of joint trajectories, utilized in the algorithm, followed by set of calculation steps, gives smooth transition between initial and final poses, as visible in the trajectory plots. The maximum velocity during transformation is limited by angular velocity threshold of servomotors located in robot body and in robot arms. The values are also limited by actual load exerted on the drives by the pedipulator structure. It is visible in Fig. 4.8 that during transformation for motion in horizontal pipes from all the selected initial poses, maximum velocities do not exceed 40 deg/s for 5-s runs that comply with the drives limits. Smooth transformations assured by boundary conditions for the polynomial interpolation of joint trajectories impose also smooth acceleration of all servomechanisms. It should be noted that the transformations are reversible, thus e.g. the trajectory from ∅242 to ∅300 mm pipe can be reversed and executed in the opposite sequence.

In Fig. 4.9, pedipulators transformation trajectories for vertical pipes are shown. Three of these trajectories require a dedicated support for the robot body, since they are used during preparation phase before placement of the robot in a vertical pipe (Fig. 4.9a–c). The transformation, depicted in Fig. 4.9d is the most important trajectory for the robot control in vertical pipes. It covers the entire extension range of the pedipulators from ∅224 to ∅270 mm diameter pipes. Proper implementation of this transformation trajectory in a control system would provide regulation of clamp force during adaptation to changing dimensions of vertical pipes. Usually, the application scenario in vertical operation would consist of the trajectories, shown in Fig. 4.9a, d for preparation and further adjustment of pedipulators extension.

The trajectories shown in Fig. 4.10 are useful during exploitation of the robot in inspection of even surfaces. Figure 4.10a presents transformation to an intermediate pose, required to precede the pose for horizontal surfaces that requires specific orientation of the motion unit. Regarding this fact, it is not possible to directly reconfigure the robot from an arbitrary pose to the horizontal surface setting.

There exist two attainable configurations of the robot with parallelly oriented tracks—a regular pose with 2-step transformation shown in Fig. 4.10b and high pose, generally unstable with drives positioned near singular configuration of the pedipulator (Fig. 4.10c). Another useful transformation is depicted in Fig. 4.10d. It is used when the robot should exit a circular cross-section pipe and continue inspection on even surface.

Initial validation of the model was performed in the MATLAB environment. The pedipulators positions were drawn for every step of the trajectory. In Fig. 4.11a, the plot of kinematic model of both pedipulators is depicted during transformation from neutral pose (all servomotors are in the middle of their operating range) to the pose that allows the robot to move in horizontal pipes with diameter ∅242 mm (Fig. 4.11b). All other transformation possibilities were visualized and verified. The pose of pedipulators for motion on horizontal surface is shown in Fig. 4.11c, whereas a pose for operation of the robot in vertical pipe is depicted in Fig. 4.11d.

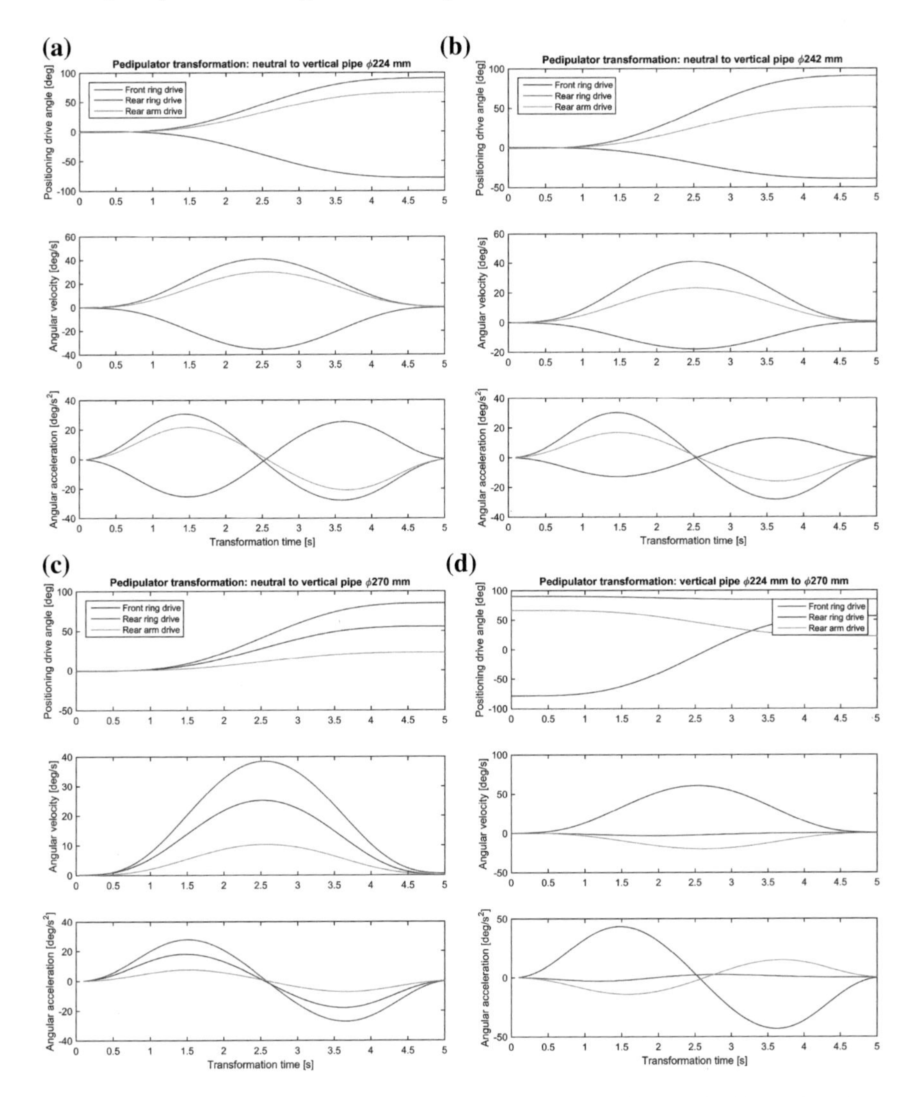

Fig. 4.9 Pedipulator transformations for vertical pipes—drives positions, velocities and accelerations: **a** from the neutral pose to a ∅224 mm pipe; **b** from the neutral pose to a ∅242 mm pipe; **c** from the neutral pose to a ∅270 mm pipe; **d** from a ∅224 to a ∅270 mm pipe

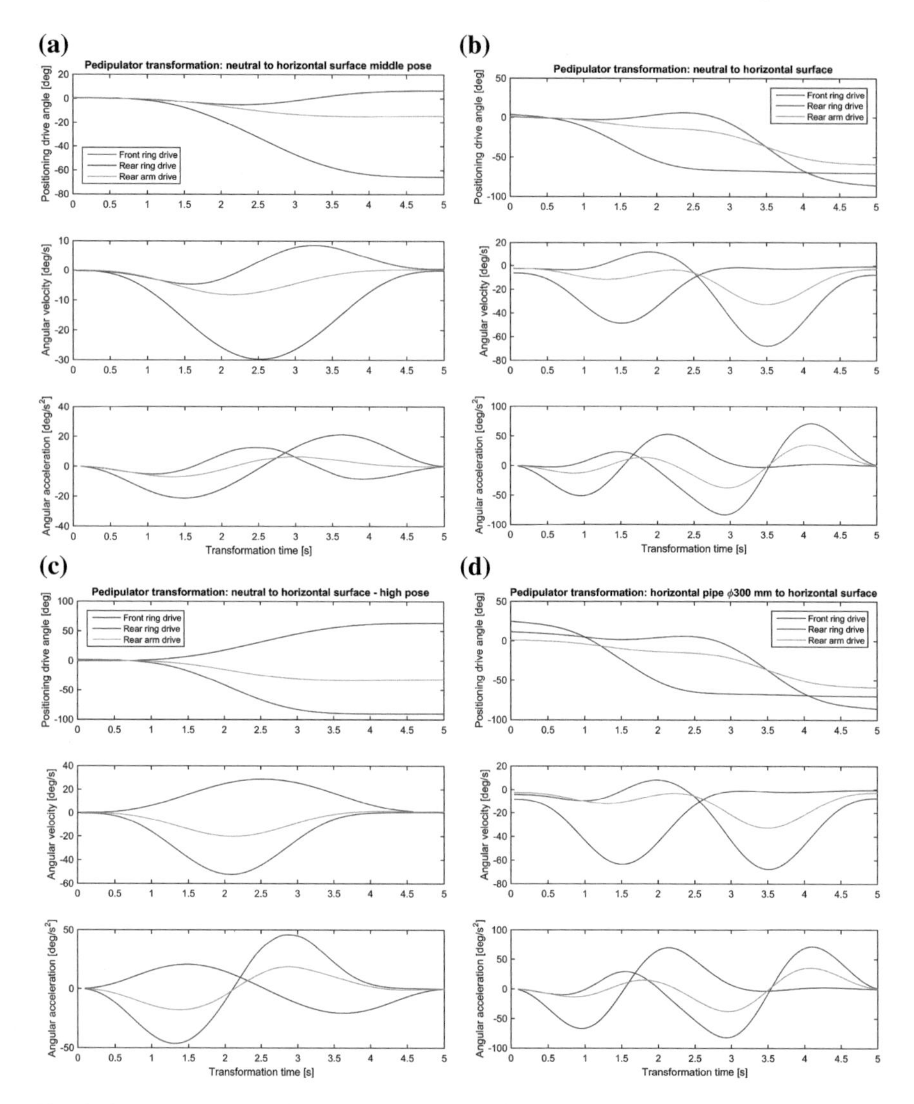

Fig. 4.10 Pedipulator transformations for horizontal surfaces—drives positions, velocities and accelerations: **a** from the neutral pose to the horizontal surface middle pose; **b** from the neutral pose to the horizontal surface—low pose; **c** from the neutral pose to the horizontal surface—high pose; **d** from a horizontal ∅300 mm pipe to the horizontal surface low pose

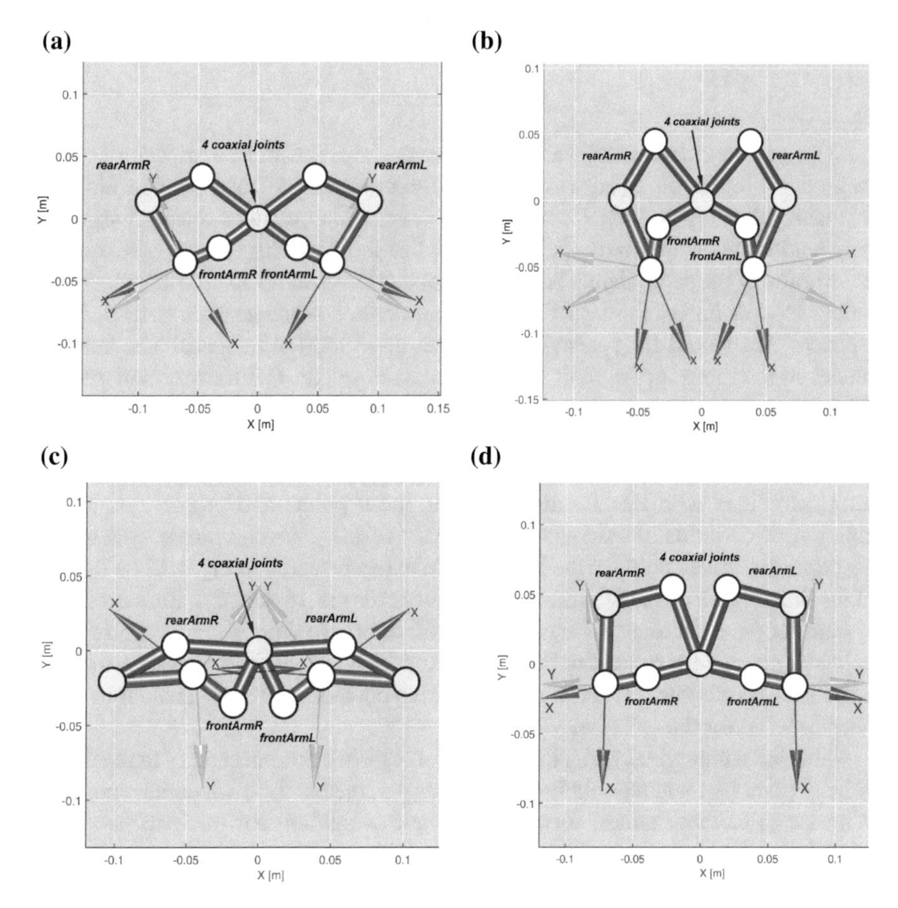

Fig. 4.11 Kinematic model of the two assembled pedipulators during transformation (white color—free joints, yellow color—actuated joints): **a** the neutral pose; **b** a pose for motion in a ∅242 mm horizontal pipe; **c** a pose for motion on a horizontal surface; **d** a pose for motion in a ∅270 mm vertical pipe

It can be concluded that with usage of the original trajectory calculation algorithm, developed by the author, it is possible to control position and orientation of the pedipulators for any desired motion in workspace. The analyzed closed kinematic chain is highly dependent on the path taken to attain particular pose, thus it is not always possible to use arbitrary start pose and expect satisfactory kinematic results for all goal poses of the pedipulators.

4.5 Pedipulators Forward Kinematics for Robot Motion in Pipes

Next step of the calculations was to visualize and assess Cartesian trajectories of the pedipulator joints, obtained with usage of the trajectory calculation algorithm. For this task, extended pedipulator model was created that includes track drive contact points with the environment. In order to realize this modeling phase, mathematical description of the pedipulators had to be augmented. It was decided to include contact points located on both sides of the track and create homogeneous transformation matrices that would fully describe the track drive motion in space. The extended model of the robot motion unit is presented in Fig. 4.12. DH transformations from the base coordinate system, located in the center of the robot body were performed with usage of rear (outer) robot arms.

Track contact points are denoted as p_{5a} and p_{5b}, since it is necessary to use five transformations with DH notation to reach these points and properly adjust orientation of coordinate systems to match axes of the base coordinate system. Full description is given only for the left track drive (represented in Fig. 4.12 on the right of the drawing, since the Z-axis is pointing forwards the robot body). Base coordinate system for the right track drive is depicted for reference with x_0^R, z_0^R axes oriented in opposite sense with respect to the x_0^L, z_0^L axes. The axes y_0^R and y_0^L are coaxial. All other transformations for the right track drive contact points are analogous to the ones denoted for the left side.

Table 4.3 summarizes DH parameters of transformations necessary to reach track contact point p_{5a}, whereas Table 4.4 specifies parameters for track contact point p_{5b}. With usage of these tables, forward kinematics equations for track contact points were derived.

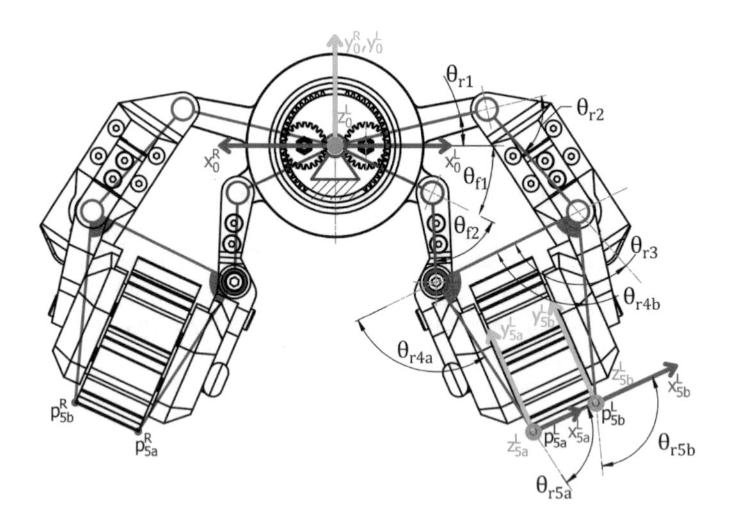

Fig. 4.12 Extended model of the pedipulators with the track contact points

Table 4.3 Denavit-Hartenberg parameters for the track contact point p_{5a}

No.	θ_i	$\mathbf{d_i}$	$\mathbf{a_i}$	α_i
1	θ_{r1}^{var}	0	a_{r1}	0
2	θ_{r2}^{var}	0	a_{r2}	0
3	θ_{r3}^{var}	0	a_{r3}	0
4	θ_{r4a}	0	a_{r4a}	0
5	θ_{r5a}	0	0	0

Table 4.4 Denavit-Hartenberg parameters for the track contact point p_{5b}

No.	θ_i	$\mathbf{d_i}$	$\mathbf{a_i}$	α_i
1	θ_{r1}^{var}	0	a_{r1}	0
2	θ_{r2}^{var}	0	a_{r2}	0
3	θ_{r3}^{var}	0	a_{r3}	0
4	θ_{r4b}	0	a_{r4b}	0
5	θ_{r5b}	0	0	0

Table 4.5 Constant parameters of the pedipulator mechanism

Parameter	a_{r1}	a_{r2}	a_{r3}	a_{r4a}	a_{r4b}	θ_{r4a}	θ_{r5a}	θ_{r4b}	θ_{r5b}
Unit	[m]					[rad]			
Value	0.05800	0.05000	0.05845	0.06469	0.06907	1.71304	1.40324	1.21737	1.89909

For calculation of forward kinematics of the pedipulator structure, constant parameters measured from the CAD model, listed in Table 4.5, were used for simplification of calculation procedures.

Transformation matrices of pedipulator forward kinematics for track contact points p_{5a}^L and p_{5b}^L are given by Eqs. (4.92) and (4.93) respectively.

$$T_{5,0}^{p_{5a}^L} = \left[\begin{matrix} -0.0253sin\left(\theta_{r1}+\theta_{r2}+\theta_{r3}\right) - 0.9997cos\left(\theta_{r1}+\theta_{r2}+\theta_{r3}\right) \\ -0.9997sin\left(\theta_{r1}+\theta_{r2}+\theta_{r3}\right) + 0.0253cos\left(\theta_{r1}+\theta_{r2}+\theta_{r3}\right) \\ 0 \\ 0 \end{matrix}\right.$$

$$\begin{matrix} -0.0253cos\left(\theta_{r1}+\theta_{r2}+\theta_{r3}\right) + 0.9997sin\left(\theta_{r1}+\theta_{r2}+\theta_{r3}\right) & 0 \\ -0.0253sin\left(\theta_{r1}+\theta_{r2}+\theta_{r3}\right) - 0.9997cos\left(\theta_{r1}+\theta_{r2}+\theta_{r3}\right) & 0 \\ 0 & 1 \\ 0 & 0 \end{matrix}$$

$$\left.\begin{matrix} 0.0493cos\left(\theta_{r1}+\theta_{r2}+\theta_{r3}\right) - 0.0640sin\left(\theta_{r1}+\theta_{r2}+\theta_{r3}\right) + 0.05cos\left(\theta_{r1}+\theta_{r2}\right) + 0.058cos\left(\theta_{r1}\right) \\ 0.0493sin\left(\theta_{r1}+\theta_{r2}+\theta_{r3}\right) + 0.0640cos\left(\theta_{r1}+\theta_{r2}+\theta_{r3}\right) + 0.05sin\left(\theta_{r1}+\theta_{r2}\right) + 0.058sin\left(\theta_{r1}\right) \\ 0 \\ 1 \end{matrix}\right] \tag{4.92}$$

$$T_{5,0}^{p_{5b}^L} = \begin{bmatrix} -0.0253sin\,(\theta_{r1}+\theta_{r2}+\theta_{r3}) - 0.9997cos\,(\theta_{r1}+\theta_{r2}+\theta_{r3}) & -0.0253cos\,(\theta_{r1}+\theta_{r2}+\theta_{r3}) + 0.9997sin\,(\theta_{r1}+\theta_{r2}+\theta_{r3}) & 0 & 0.0824cos\,(\theta_{r1}+\theta_{r2}+\theta_{r3}) - 0.0648sin\,(\theta_{r1}+\theta_{r2}+\theta_{r3}) + 0.05cos\,(\theta_{r1}+\theta_{r2}) + 0.058cos\,(\theta_{r1}) \\ -0.9997sin\,(\theta_{r1}+\theta_{r2}+\theta_{r3}) + 0.0253cos\,(\theta_{r1}+\theta_{r2}+\theta_{r3}) & -0.0253sin\,(\theta_{r1}+\theta_{r2}+\theta_{r3}) - 0.9997cos\,(\theta_{r1}+\theta_{r2}+\theta_{r3}) & 0 & 0.0824sin\,(\theta_{r1}+\theta_{r2}+\theta_{r3}) + 0.0648cos\,(\theta_{r1}+\theta_{r2}+\theta_{r3}) + 0.05sin\,(\theta_{r1}+\theta_{r2}) + 0.058sin\,(\theta_{r1}) \\ 0 & 0 & 1 & 0 \\ 0 & 0 & 0 & 1 \end{bmatrix} \quad (4.93)$$

With usage of transformation data specified above and trajectories for particular transformations, derived in Sect. 4.4, it is possible to determine robot position in pipelines. It was assumed that robot pedipulators perform synchronized motion, thus pose of pedipulator on one side of the robot is symmetrical with respect to the one on the other side. With this assumption, by selecting three track contact points, it is possible to describe a circle on these points that may represent simplified pipe model. The fourth contact point would be in this case also located on the constructed circle. Set of equations that can be used to find center point coordinates and radius of the circumscribed circle is given by Eqs. (4.94).

$$\begin{cases} (x_1 - x_C)^2 + (y_1 - y_C)^2 = r_C^2 \\ (x_2 - x_C)^2 + (y_2 - y_C)^2 = r_C^2 \\ (x_3 - x_C)^2 + (y_3 - y_C)^2 = r_C^2 \end{cases} \quad (4.94)$$

where: $x_1, x_2, x_3,\ y_1, y_2, y_3$*—coordinates of track contact points;* x_C, y_C*—coordinates of circle center point;* r_C*—radius of circle.*

Analytical solution of this set of equations gives results for arbitrarily selected three points on a plane (Eqs. 4.95–4.97):

$$x_C = \frac{(y_C - y_1)^2 - (y_C - y_2)^2 + x_1^2 - x_2^2}{2x_1 - 2x_2} \quad (4.95)$$

$$y_C = \frac{(x_2 - x_3)x_1^2 + (-x_2^2 + x_3^2 - y_2^2 + y_3^2)x_1}{(2y_3 - 2y_2)x_1 + (2x_2y_1 - 2x_3y_1 - 2x_2y_3 + 2x_3y_2)} + \frac{\left(x_2^2x_3 - x_2x_3^2 + x_2y_1^2 - x_2y_3^2 - x_3y_1^2 + x_3y_2^2\right)}{(2y_3 - 2y_2)\,x_1 + (2x_2y_1 - 2x_3y_1 - 2x_2y_3 + 2x_3y_2)} \quad (4.96)$$

$$r_C = \sqrt{x_C^2 - 2x_Cx_1 + x_1^2 + y_C^2 - 2y_Cy_1 + y_1^2} \quad (4.97)$$

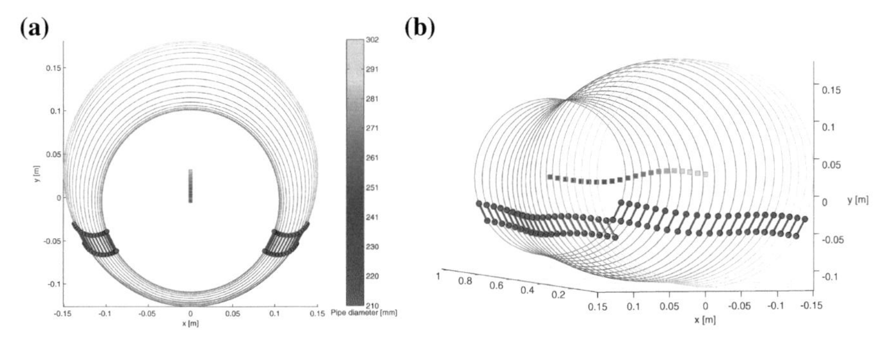

Fig. 4.13 Visualization of the robot transformation trajectory in a horizontal pipe with diameter change from ∅300 to ∅210 mm: **a** 2D view; **b** 3D view

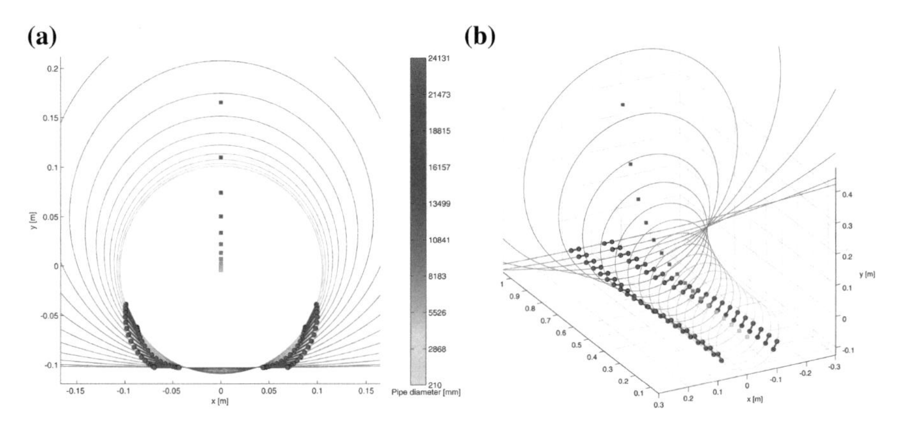

Fig. 4.14 Visualization of the robot transformation trajectory from a ∅210 mm horizontal pipe to a horizontal surface: **a** 2D view; **b** 3D view

By using numerical data for pedipulator joints, obtained from the trajectory calculation algorithm, visualizations of transformation trajectories in circular cross-section pipes were obtained. The results for transformation in a horizontal pipe are presented in Fig. 4.13. Increase of z coordinates for consecutive steps was selected to match visualization requirements. Changes of pipe cross-section center point positions are depicted, since the robot body center point is constrained at (0, 0, z) coordinates, where the base coordinate system is located. Traces of the track contact points are presented. Outline of pipe profile at consecutive transformation stages is pictured.

Analogous computation procedure was applied to robot pose transformation from horizontal pipe with circular cross-section to horizontal surface. The visualization of this pedipulator trajectory is shown in Fig. 4.14.

Due to the fact that the robot is also intended for operation in vertical pipes, visualization of parallel track extension was prepared. Figure 4.15 depicts available extension limits for the designed robot structure. We may observe that track contact points are located on opposite sides of pipe cross-section.

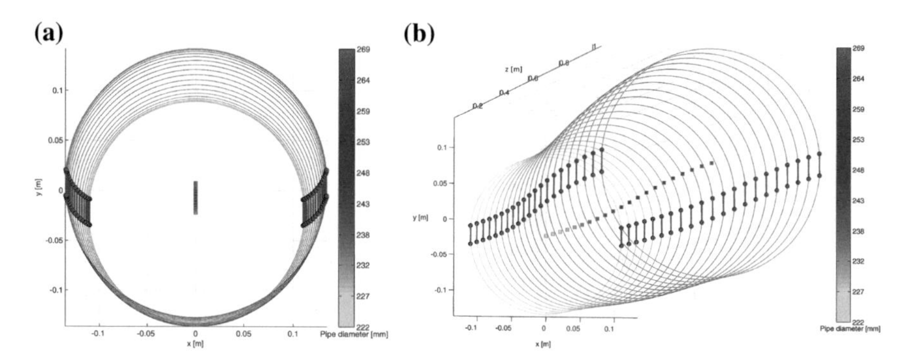

Fig. 4.15 Visualization of the robot transformation trajectory in a vertical pipewith diameter change from ∅224 to ∅270 mm: **a** 2D view; **b** 3D view

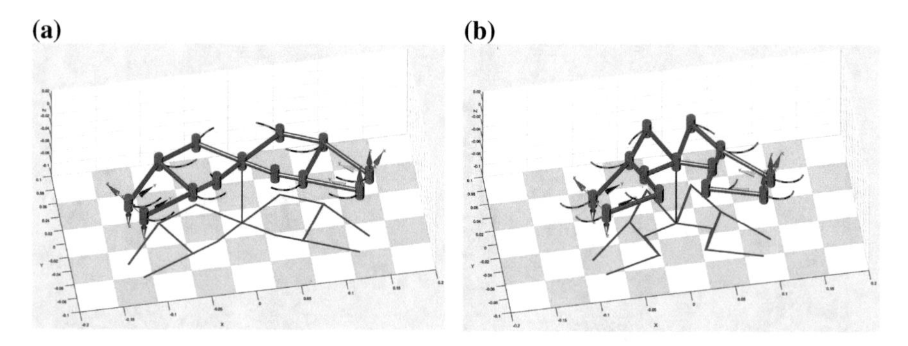

Fig. 4.16 Pedipulators transformation trajectory from the neutral pose to a ∅210 mm horizontal pipe: **a** initial pose; **b** final pose

As we may observe, visualizations of the robot transformation trajectories in pipes theoretically prove correct design of the calculation algorithm for different motion scenarios.

The transformation trajectories for the augmented 2D model of the pedipulators with track contact points were simulated in MATLAB. As in the previous cases, visualizations of the trajectories were prepared for the most probable operation scenarios. All characteristic points were marked with red cylinders. Their trajectories are shown using multi-colored curves. Coordinate system of track contact points are attached to the structure, according to DH notation and all dimensions are expressed in meters. Shadows of the pedipulators outline are projected on the base work plane, whilst the model is elevated for better visual representation of the plot. Figure 4.16 depicts the pedipulators model transformation from neutral pose to the pose for motion in horizontal pipes with diameter ∅210.

Figure 4.17 shows transformation from a pose dedicated for operation in horizontal pipes to a pose of the robot that allows motion on horizontal surfaces. It is necessary to attain an intermediate pose during the transformation (Fig. 4.17b) and then generate trajectory to the pose for horizontal and inclined surfaces, as shown in Fig. 4.17c, d.

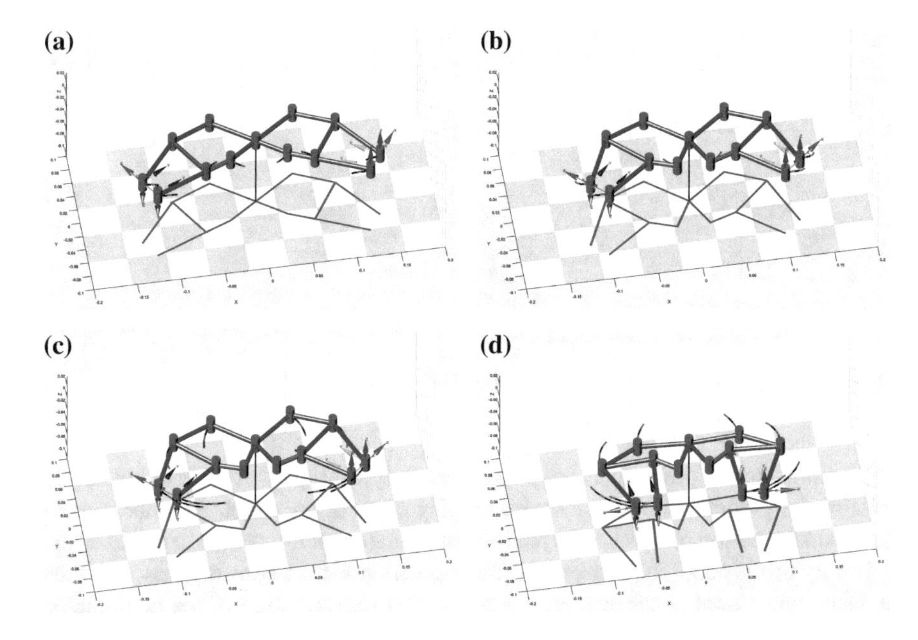

Fig. 4.17 Pedipulators transformation trajectories from a ∅300 mm horizontal pipe to a horizontal surface, through an intermediate pose: **a** step 1; **b** step 2; **c** step 3; **d** step 4

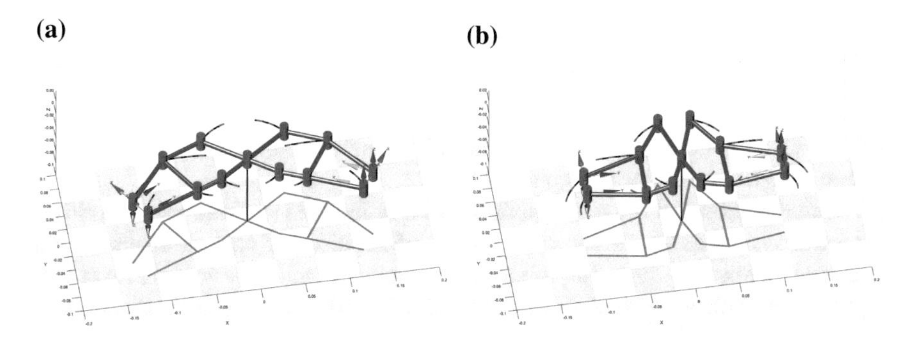

Fig. 4.18 Pedipulators transformation trajectory from the neutral pose to a ∅224 mm vertical pipe: **a** initial pose; **b** final pose

To allow the robot to move in vertical pipes, it is necessary to start transformation in neutral pose and then transform to the pose for minimum diameter of a vertical pipe as presented in Fig. 4.18.

Adaptation of the robot structure for inspection of vertical pipes is performed by usage of the trajectory shown in Fig. 4.19. It is used for initialization of a robot run in a vertical pipe and then appropriate final pose for desired pipe diameter is attained. The trajectory can be used in a control system to automatically adjust extension force exerted by track drive modules on pipe walls to maintain stable traction in pipes with imperfections or varying diameter, due to accumulation of sediments.

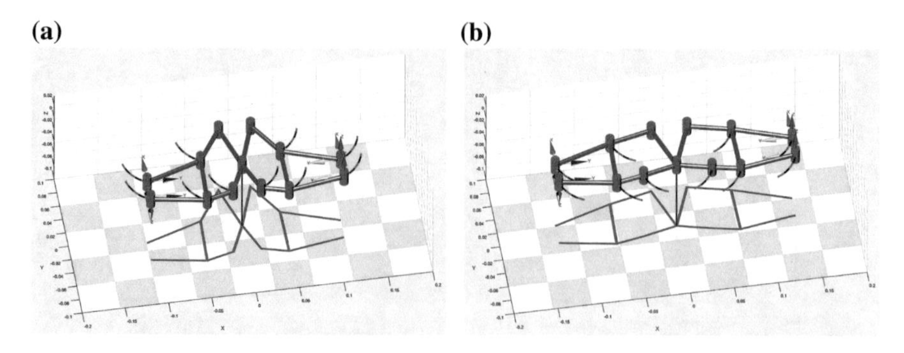

Fig. 4.19 Pedipulators transformation trajectories in vertical pipes: **a** ⌀224 mm; **b** ⌀270 mm

The trajectories presented above represent only examples of inverse kinematics solution for predefined poses, most commonly used for the robot. By selection of a goal transformation matrix for the pedipulator structure, it is possible to calculate any trajectory that is possible to execute by the robot motion unit. Nevertheless, care must be taken when goal poses are set, due to the fact that structure of the pedipulators closed kinematic chain may introduce difficulties in finding valid solution for inverse kinematics.

4.6 Summary

Mathematical modeling presented in this chapter includes several aspects of control system design for the pipe inspection robot.

Kinematic model of the robot motion on flat or inclined surfaces may be utilized in control system as an additional source of information for pose estimation in the workspace. However, since the robot motion unit is a skid-steering platform with highly deformable track treads, other sources are needed for proper localization.

The dynamic model is only applicable to flat and inclined surfaces and it is highly dependent on environmental conditions, thus it should be treated as a reference in initial design phase of control system. Intended operation scenarios of the robot include mainly pipelines, where supervised forward and backward motion control is sufficient for execution of video inspection tasks.

The most important issue, indispensable for operation of the robot, is the modeling of pedipulator transformation and calculation of track drive trajectories. The original algorithm for solving inverse kinematics problem of the closed kinematic chain plays key role in the control system of the robot. It enables adaptation of the drive unit to different pipe sizes and shapes, flat surfaces, rough terrain and vertically oriented pipelines.

References

1. Ciszewski M, Waclawski M, Buratowski T, Giergiel M, Kurc K. Design, modelling and laboratory testing of a pipe inspection robot. Arch Mech Eng. 2015;62(3):395–408.
2. Trojnacki M. Modelowanie i symulacja ruchu mobilnego robota trzykołowego z napdem na przednie koła z uwzględnieniem poślizgu kół jezdnych. Modelowanie Inżynierskie. 2011;10(41):411–20.
3. ŻylskiW. Kinematyka i dynamikamobilnych robotów kołowych.OficynaWydawnicza Politechniki Rzeszowskiej; 1996.
4. Burdziński Z. Teoria ruchu pojazdu gaąsienicowego.Warszawa:Wydawnictwo Komunikacji i Łączności; 1972.
5. Ciszewski M, Buratowski T, Giergiel M, Kurc K, Małka P. The pipes mobile inspection robots. Diagnostyka. 2012;3(63):9–15.
6. Giergiel M, Buratowski T, Małka P, Kurc K. The mathematical description of the robot for the tank inspection. In: Mechan Mech Eng. 2011;15.4:53–62.
7. Wong JY. Terramechanics and Off-road vehicle engineering. In: Oxford: Elsevier, Butterworth-Heinemann; 2010. p. 155–76.
8. Ciszewski M, Buratowski T, Giergiel M, Malka P, Kurc K. Virtual prototyping, design and analysis of an in-pipe inspection mobile robot. J Theor Appl Mech. 2014;52(2):417–29.
9. Engel Z. Giergiel J. Dynamika. Mechanika techniczna, 2nd ed. Kraków:Wydawnictwa AGH; 1998.
10. Inuktun Services Ltd. Inuktun crawler vehicles. 2015. http://www.inuktun.com/crawler-vehicles. Accessed 25 Oct 2015. Chapter 4. Mathematical modeling of the robot 68
11. Giergiel M, Hendzel Z, Żylski W. Modelowanie i sterowanie mobilnych robotów ko?owych. Warszawa: Wydawnictwo Naukowe PWN; 2013.
12. Frączek J, Wojtyra M. Kinematyka układów wieloczłonowych.Warszawa:Wydawnictwo Naukowo-Techniczne; 2008.
13. Kozłowski K, Dutkiewicz P, Wróblewski W. Modelowanie i sterowanie robotów. 1st ed. Warszawa: Wydawnictwo Naukowe PWN; 2012.
14. Corke P. Robotics, vision and control: fundamental algorithms inMATLAB, vol. 3. Springer Science & Business Media; 2011.
15. Sybilska AM. Porównaniemetodywykorzystującej proste i odwrotne zadanie kinematyki oraz jakobian do sterowania manipulatorem. MA thesis. PolitechnikaWarszawska; 2007.
16. Wu CH, Young KY. An efficient solution of a differential inverse kinematics problem for wrist-partitioned robots. IEEE Trans Robot Autom. 1990;6(1):117–23.
17. Dulba I, Opałka M. A comparison of Jacobian-based methods of inverse kinematics for serial robot manipulators. Int J Appl Math Comput Sci. 2013;23(2):373–82.
18. Buss SRS. Introduction to inverse kinematics with jacobian transpose, pseudoinverse and damped least squares methods. In: 132.4. University of California;2009. p. 1–19.
19. Spong MW, Hutchinson SM V. Robot modeling and control. In: Control 141.1, 2006. p. 419.
20. Barinka L, Berka R. Inverse kinematics—basic methods (Report). Technical report: Czech Technical University; 2002.
21. Biagiotti L, Melchiorri C. Trajectory planning for automatic machines and robots. 1st ed. Berlin, Heidelberg: Springer; 2008.
22. Maempel J. Koch T. Koehring S. Obermaier A.Witte H. Concept of a modular climbing robot. In: 2009 IEEE symposium on industrial electronics & applications 2; 2009. p. 789–94.
23. Giergiel J, Giergiel M, Buratowski T, Ciszewski M. Mechanizm pedipulatora do ustawiania pozycji modułu napdowego, zw?aszcza robota mobilnego. PL2238752016.
24. Corke P. Robotics Toolbox for MATLAB—Release 9.10 manual. 2015. http://www.petercorke.com/robot. Accessed 10 Jun 2016.
25. Corke P. Robotics Toolbox. 2016. http://petercorke.com/Robotics_Toolbox.html. Accessed 10 Feb 2016.

Chapter 5
Simulations of the Robot Adaptation and Motion in Various Environments

5.1 Introduction

To validate the mathematical models and the original trajectory calculation algorithm of the robot pedipulators, it was indispensable to use a simulation environment. Previously described theoretical experiments featured only 2D representation of the robot motion unit. Application of a specially prepared simulation environment, allowed to perform extensive tests of the robot control system interaction with the 3D model, designed in the Autodesk Inventor CAD software.

At present, there exists large number of robotic simulators. They differ in functionality, extensibility, communication features and interface. According to [1], Gazebo, Open Dynamics Engine (ODE), ARGoS and Virtual Robot Experimental Platform (V-REP) simulation software are the most popular. The ODE, Bullet and ARGoS are library packages that can be used for simulation, whereas Gazebo and V-REP simulators have graphical user interface. The built-in dynamic simulation environment in MATLAB is also capable of simulations of robot complex behavior [2]. The most adapted MATLAB tool for dynamic motion simulation is the Simulink SimMechanics module that allows import and easy manipulation of 3D models [5]. On the other hand, V-REP software is a very versatile and customizable simulation platform that can be easily connected with remote applications such as MATLAB and can be used for various research problems [8]. It can be run on Linux, Windows and Mac operating systems and its complete work environment provides easy file transfer between different users. When comparing V-REP and Gazebo, the first one is a more intuitive and user-friendly simulator, and has more features whilst Gazebo is an open-source solution, more integrated into ROS (Robot Operating System) framework that is widely used in robot control applications. However, in order to match V-REP functionalities, a considerable amount of custom extensions of Gazebo functionalities is required [3]. Thus, V-REP was chosen as the main simulator for validation of the robot control system and dynamic behavior during motion in various environments.

M. Ciszewski et al., *Modeling and Control of a Tracked Mobile Robot for Pipeline Inspection*, Mechanisms and Machine Science 82, https://doi.org/10.1007/978-3-030-42715-3_5

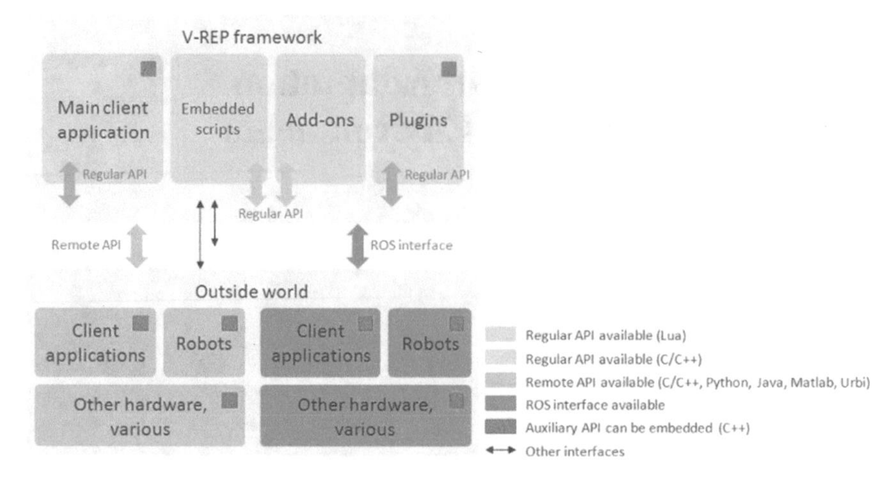

Fig. 5.1 Overview of the V-REP application programming interface [7]

Due to the fact, that the robot control algorithms were developed in MATLAB, a remote API was used to connect to V-REP simulator. Remote APIs, independently of programming language chosen, offer four operation modes, each of them designed to execute different tasks (Fig. 5.1). First is a blocking function call that causes API client to wait for response of simulator. Due to communication delays between client and server, the application is blocked until whole inquiry is completed. It should be used only when data obtained from a query is required directly after performing it. Second type is a non-blocking function call that could be used only for sending data to the simulator, as it does not wait for server response. Third type is data streaming that could be considered as a kind of subscription, where a client sends a message once and a server sends reply regularly to the client. Data is stored in a buffer until the client reads it or buffer limit is reached. The last mode is synchronous operation. In contrast to previous modes, this one enables to synchronize each simulation step with the API client. If simulation is started in this mode, a server holds execution of each simulation step, until a trigger from a client is sent. It is also worth mentioning that sending of a trigger is always executed as in blocking mode, thus synchronous operation is usually the slowest mode of communication.

5.2 Robot Models in MATLAB and V-REP Software

The first step of simulations was to create a properly defined model in V-REP that satisfies all mechanical constraints and can be simulated in a way resembling real prototype operation. The robot 3D model designed in CAD software was imported in the V-REP environment. Next, the model structure was defined in a way that allowed creation of joints, motors and ensured interaction with environment. To improve

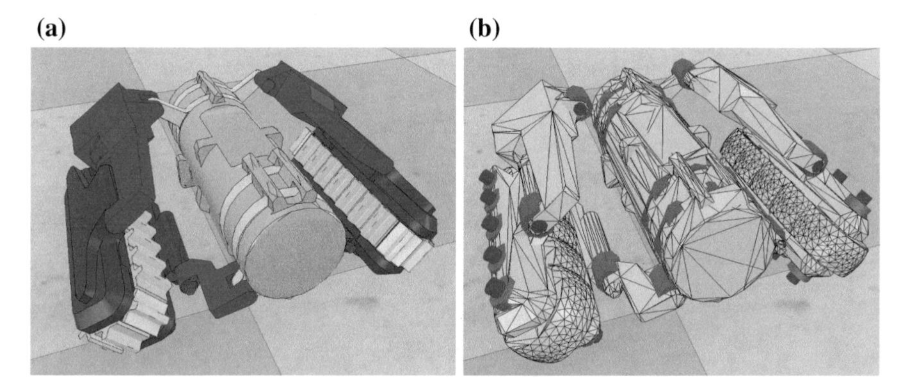

Fig. 5.2 V-REP model of the pipe inspection robot: **a** visible bodies; **b** dynamically enabled bodies (green color) with joints (blue-orange color)

calculation efficiency, the model was composed of visible bodies and dynamically enabled bodies that were simplified using convex hull decomposition [7]. Overview of the robot model is presented in Fig. 5.2. To improve visual aspects of the simulation, the tracks were modeled using separate segments that move on defined paths (Fig. 5.2a). In contrast, for dynamic model, each track was represented by five spherical rollers [4]. This approach proved to be the most stable for simulations (Fig. 5.2b), providing maximally 10 contact points with the environment. Natural representation of a track as a set of segments is computationally costly and may lead to simulation instability due to large number of rapidly alternating contact points. The pedipulator mechanisms were modeled using rotary joints that were set as actuated with position control or free, according to the structure of the mechanism.

Model hierarchy of the pipe inspection robot is presented in Fig. 5.3. All elements are named according to their functions. The elements with *_dyn* suffix are dynamically enabled bodies that are analyzed during simulation. Dynamic bodies have their visual substitutes, placed in corresponding subtrees. We can also distinguish other elements such as cameras, joints and dummies (points with assigned orientation). All joints that are attached to dynamic bodies are by default dynamically enabled. For closure of kinematic chains, dummy elements are used. They are located in the model tree of both elements that have to be connected and can serve as links, indicated by blue arrows in Fig. 5.3. The model building approach in V-REP is based on hierarchy, so to ensure reasonable simulation results, it is necessary to properly arrange model elements.

V-REP provides support for 4 dynamic calculation engines: Bullet physics library, ODE, Vortex and Newton that have different functionalities and can be chosen depending on the type of simulation [7]. The Bullet physics library is an open source physics engine featuring 3D collision detection, rigid body dynamics, and soft body dynamics (not available in V REP). It proved to be the most stable for the robot simulations. Complex contact calculations between the tracks substituted by rollers and pipe surfaces were realized in a predictable manner. Simulation parameters, listed in Table 5.1 were adjusted to obtain the best performance to precision ratio.

Fig. 5.3 Hierarchical structure of the pipe inspection robot model created in V-REP

Table 5.1 Parameters of the V-REP simulations

Parameter	Value
Dynamics engine	Bullet physics library
Simulation time step for horizontal pipe and horizontal surface	0.05 [s]
Simulation time step for vertical pipe	0.01 [s]
Collision detection	Enabled
Calculation accuracy	Very accurate

The simulation procedure involves control commands that cannot be directly executed in V-REP environment with usage of built-in scripts. Therefore, the robot model created in V-REP was linked with MATLAB environment using a remote API link. The link allows running simulations directly from MATLAB, with simultaneous data flow between both programs [6]. Simulations can be run synchronously, with each calculation step, executed concurrently in both programs. It can be referred to as soft real-time execution, because depending on model complexity and used hardware it might not be possible to run simulations in real-time. The mathematical model created in MATLAB was used to control positions of the pedipulators' servomotors and also for velocity control of the track drive modules. For creation of the control procedure, it was necessary to establish V-REP communication handles and signals that were used to pass MATLAB calculations to V-REP environment. The calculations of pedipulators poses were integrated with the control procedure to resemble operation a prototype in real life conditions. MATLAB functions were linked with a Simulink model for seamless robot control. The Simulink model is presented in Fig. 5.4, featuring communication interface with a joystick, track velocity control with signals received from the joystick, integration of the mathematical model for robot simulations in vertical pipe and dedicated functions that realize communication with the V-REP simulator. The Logitech F710 joystick was used for controlling of robot motion. A custom Simulink block was developed, based on joystick hardware interface, prepared as MATLAB function. The joystick is equipped with analog and digital inputs, thus allowing for easy and individual control of track velocities, reconfiguration procedures and other robot functionalities. The velocities can be controlled in two ways, accustomed for different work scenarios. The first method is dedicated for straight pipe segments and comprises of linear velocity setting by analog buttons for forward and backward track motion. The second method enables individual control of track velocity and direction of motion by means of two multidirectional sticks. Communication function block for linking V-REP simulator provides transfer of control data directly to specific drives in the robot model (represented on the right side of Fig. 5.4). Blocks that provide functionalities for simulation in vertical pipes are located at the bottom of the scheme, depicted in Fig. 5.4. After attaining a specific pose for vertical pipe, it is possible to manually adapt the robot pose for vertical pipe with diameters ranging between ⌀224 and ⌀270 mm. The adjustment

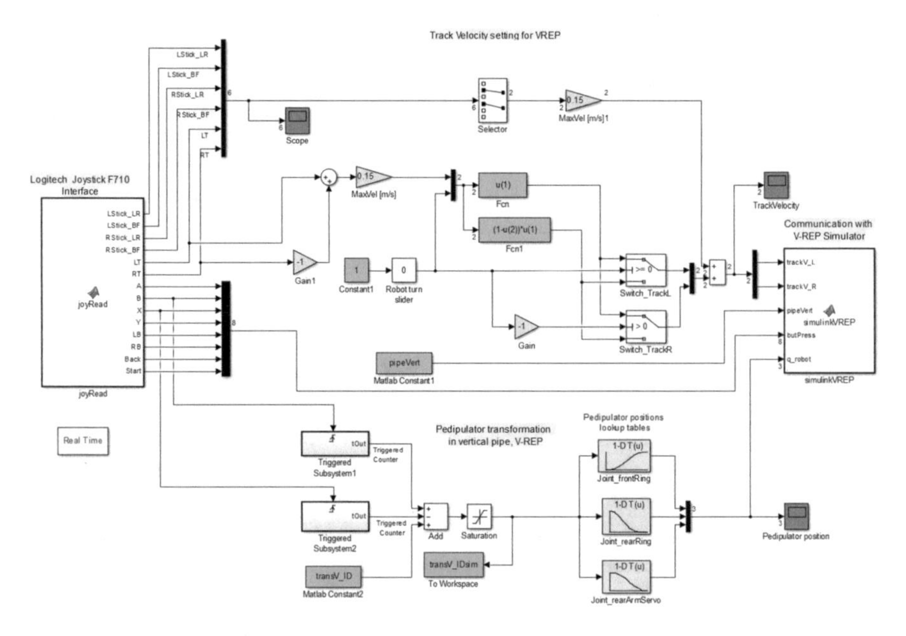

Fig. 5.4 The MATLAB/Simulink model used for hardware interaction and communication with the V-REP simulator

can be realized in 50 steps during robot simulation with usage of the joystick buttons, therefore a satisfactory extension force can be obtained for particular pipe size.

Comprehensive simulation scenarios were prepared to thoroughly investigate correctness of robot transformation trajectories and operation in varying environments. Initially, transformation trajectories of the pedipulators were verified by simple V-REP simulations that involved only the robot model located on a support, next robot motion was tested in various environments. In order to check operation of the robot in complex pipe structures, dedicated scenes were prepared.

5.3 Horizontal Pipe Run

Having completed validation of all mathematical models of the robot, a scene with a horizontal pipe run was prepared (Fig. 5.5a). The pipe run contained the following segments: straight ∅300 mm pipe, 90° bend in ∅300 mm pipe followed by straight segment, reducer from ∅300 to ∅242 mm, straight ∅242 mm segment followed by 30° bend and straight section. A co-simulation in MATLAB, Simulink and V-REP was run to check if the robot would be able to traverse the pipe assembly.

Reconfiguration of the pedipulators was performed automatically, based on the mathematical model, whereas motion of the robot in pipe was realized by manual control with usage of the joystick connected to the MATLAB/Simulink model. In

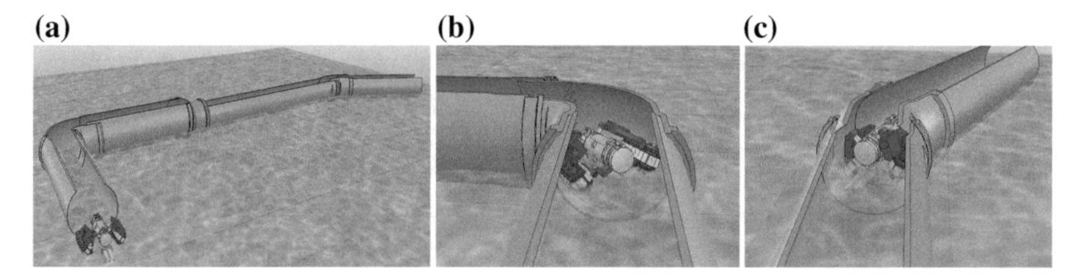

Fig. 5.5 V-REP model of the pipe inspection robot: **a** simulation environment of a horizontal pipe run; **b** the robot negotiating a 90° bend in a ∅300 mm pipe; **c** the robot negotiating a 30° bend in a ∅242 mm pipe

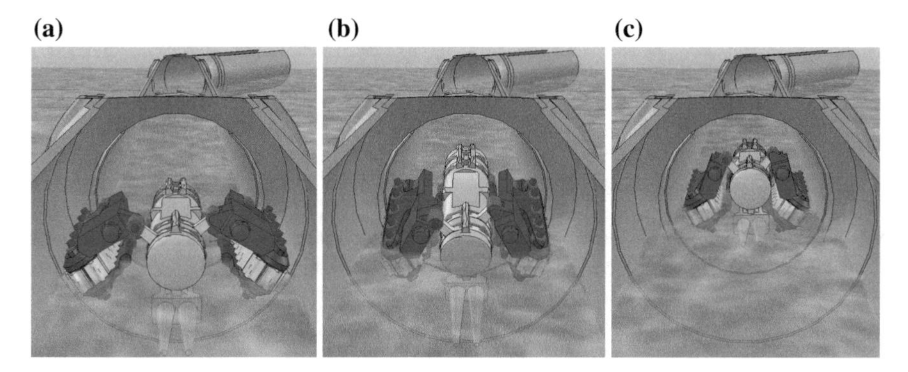

Fig. 5.6 V-REP simulation of the robot reconfiguration in a reducer from a DN315 pipe to a DN250 pipe: **a** the robot in a ∅300 mm pipe segment; **b** the robot in the reducer; **c** the robot after reconfiguration in a ∅242 mm pipe

Fig. 5.5b we can see the robot during motion in 90° bend of a ∅300 mm pipe, whereas Fig. 5.5c depicts the robot in 30° bend of a ∅242 mm pipe. The most difficult simulation phase, when the robot passes a reducer is shown in Fig. 5.6. The process of driving through narrowing or widening pipe section involves simultaneous control of track velocity and reconfiguration of both pedipulators. It was achieved by optimal matching of track velocity and pedipulators actuation speed. In Fig. 5.6a, the robot is shown in a ∅300 mm pipe before entering the reducer, Fig. 5.6b depicts intermediate phase of reconfiguration and Fig. 5.6c depicts the robot after entering to a ∅242 mm pipe segment. It should be noted that according to the simulation, it might be impossible to negotiate such pipe connectors without synchronized operation of all robot drives. Friction forces between the tracks and the pipe surface are also important factors and for pipes filled with liquids, causing drastic decrease of friction, a reducer could become a problematic geometric obstacle.

Results of the simulation performed in the horizontal pipe run are presented in Fig. 5.7. We can see that the mathematical model trajectories and simulation results from V-REP coincide well and the absolute error does not exceed 0.6° for the analyzed cases, proving proper model definition. Discrepancies in simulation results that occur in V-REP model at the beginning of transformation phase (Fig. 5.7a) are caused by

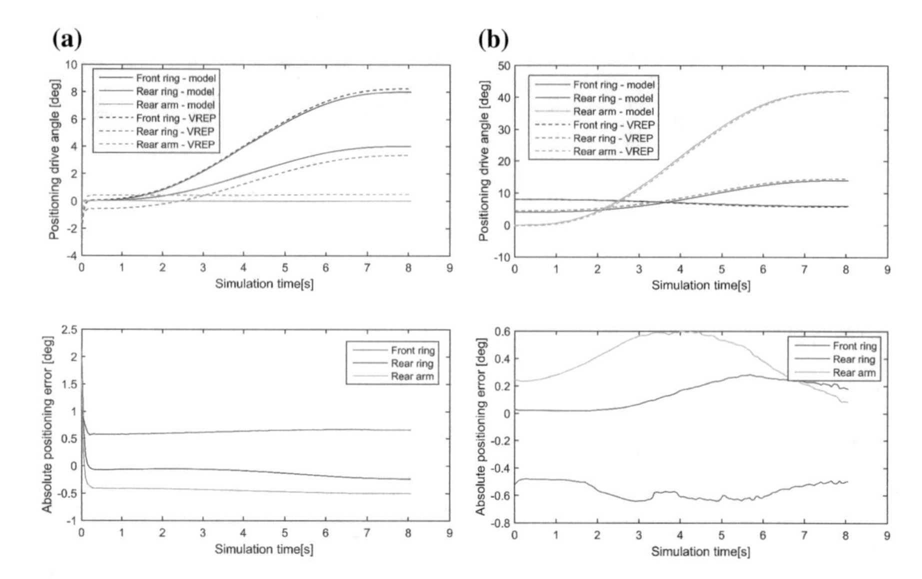

Fig. 5.7 V-REP and MATLAB simulations—transformation results comparison and absolute positioning error: **a** from the neutral pose to a ∅300 mm horizontal pipe; **b** from a ∅300 mm pipe to a ∅242 mm pipe

initialization of simulation run that takes approximately 0.2 s to stabilize. Initial error values reaching 2.2° are shortly decreased and position errors maintain lower values throughout the rest of the simulation.

With usage of the simulation it was checked that the robot operation would be possible in horizontal pipes of varying diameters that include reducers and bends.

5.4 Motion on Even and Rough Surfaces

Another type of robot operation environment can consist of horizontal surfaces or rough terrain that may be encountered during inspection of hardly reachable places, floor of liquid storage tanks and water wells. With the adjustable chassis, the robot may be utilized as a versatile inspection tool. For the purpose of motion validation, a designated scene with flat surface and rough terrain was prepared in V-REP. The previously elaborated mathematical models were checked for reconfiguration to motion pose dedicated for even surfaces. In Fig. 5.8a, the robot in neutral pose located on an elevated support is depicted. The support is utilized to reconfigure the pedipulators before placing the robot in particular environment. After reconfiguration, the support is removed and the robot can be simulated normally. In Fig. 5.8b we can observe the robot during motion on a horizontal surface, approaching rough terrain.

Figure 5.9 present plots of pedipulators positioning drives angles during reconfiguration, shown in Fig. 5.8. This reconfiguration trajectory is the most complicated case

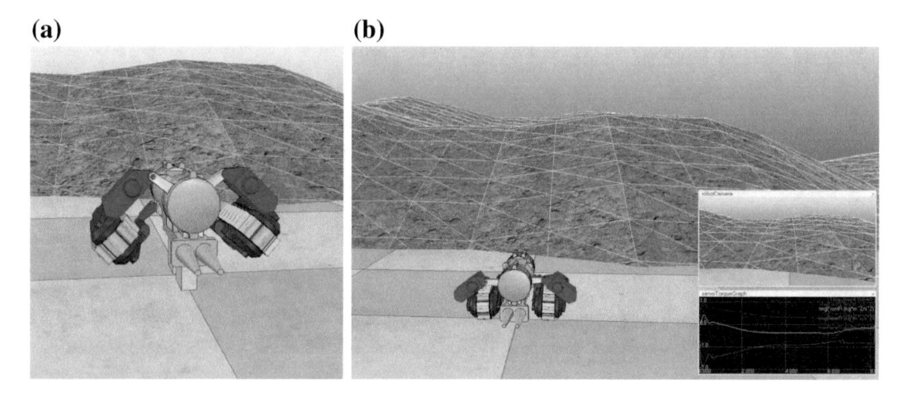

Fig. 5.8 V-REP simulation of the robot reconfiguration: **a** the robot in the neutral pose located on a support; **b** the robot on a horizontal surface after reconfiguration to the horizontal high pose

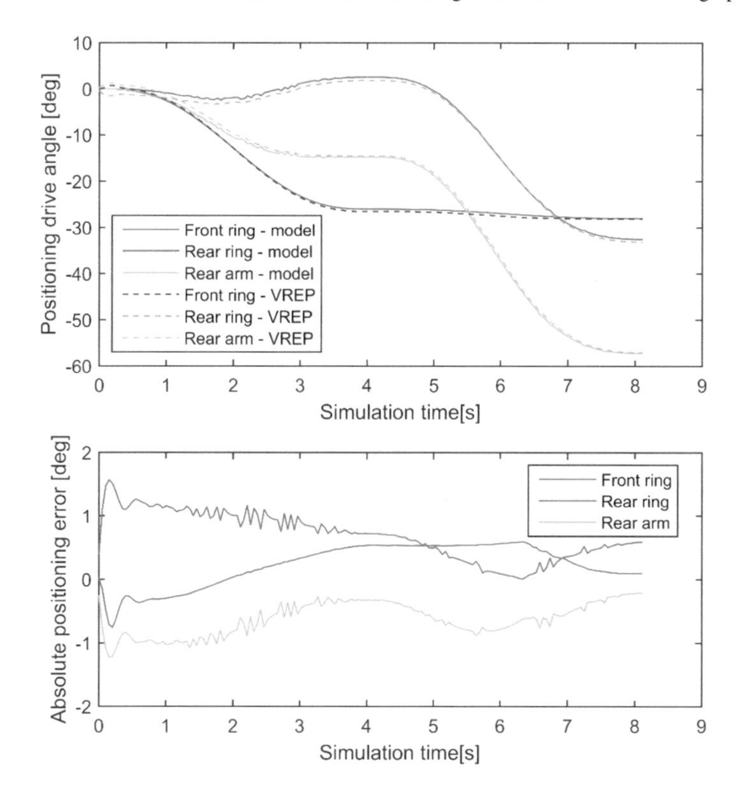

Fig. 5.9 V-REP and MATLAB simulations—transformation results comparison and absolute positioning errors: from the neutral pose to the horizontal high pose

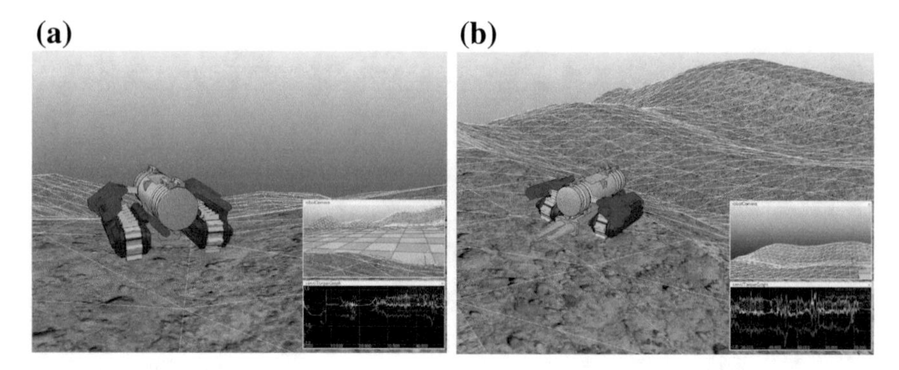

Fig. 5.10 V-REP simulation of the robot motion in rough terrain: **a** the robot after driving uphill; **b** downhill motion of the model

that comprises of a predefined, intermediate pose that allows the inverse kinematics model to solve the equations, whilst preserving condition of the closed kinematic chain. The lower plot presented in Fig. 5.8 indicates slight fluctuations in absolute positioning error for *Rear ring* and *Rear arm joints* that are caused by vibrations of the simulated mechanism.

General motion on rough terrain is difficult to analyze thoroughly due to composition discrepancies of various ground types. Simulations presented in Fig. 5.10 were prepared for an arbitrarily generated rough terrain model with slope inclination not exceeding 20°. View from camera installed on the robot is visible in upper subfigures of Fig. 5.10a, b, whereas torque plot of pedipulator positioning servomotors is depicted in lower subfigures, respectively.

The robot manual joystick control proved that it can move on rough surface with changing inclination, however the results should be regarded only as coarse approximation of real operation. For exact results on robot mobility, prototype tests would be the best source of information.

5.5 Vertical Pipe Run

The most difficult robot motion scenario involves operation in vertical pipes by means of track extension and clamping to pipe walls. A scene with altered environmental conditions (gravity vector direction along pipe axis and sense opposite to robot front panel) was prepared, depicted in Fig. 5.11. It included a grounded pipe with changing diameter in the robot's range of operation (⌀224 ÷ 270 mm). The robot was positioned on a support that allowed transformation from neutral pose to vertical pipe pose.

In Fig. 5.12a, the robot is shown in a ⌀270 mm vertical pipe segment. In order to move vertically, it was necessary to first reconfigure the robot to smaller diameter pipe and after that insert it on the support to the vertical pipe segment. By manual

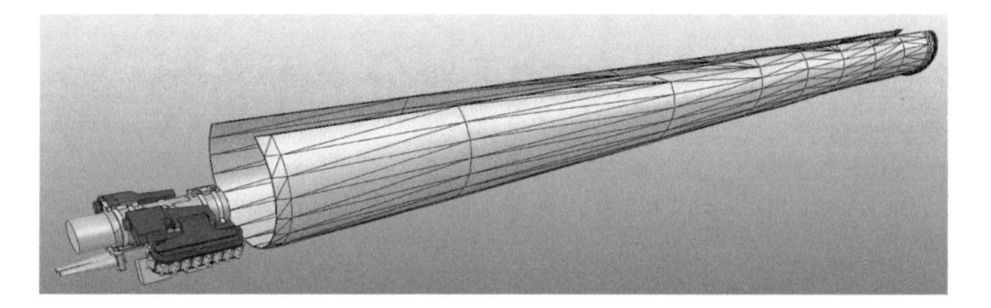

Fig. 5.11 V-REP model of the pipe inspection robot: simulation scene of a vertical pipe run

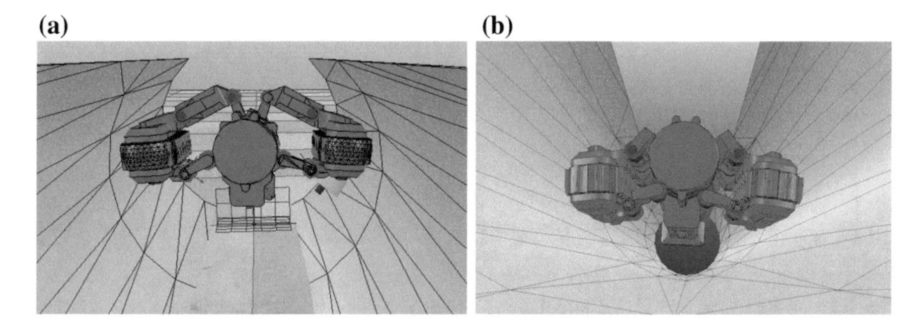

Fig. 5.12 V-REP model of the pipe inspection robot during motion in vertical pipes: **a** ∅270 mm; **b** ∅224 mm

adjustment of track extension against pipe walls, when sufficient friction force was attained, the robot model started moving vertically. In the Fig. 5.12a, the actual dynamic model of elements in contact is shown. Since the track contact surfaces are modeled as 5 cylindrical rollers, located one after another, it can be noted that maximum theoretical number of contact points between the pipe and tracks is 10. It is significantly smaller contact area than for an elastomeric track, utilized in the robot prototype that can undergo elastic deformations and adapt to pipe shape. Consequently, it can be predicted that operation of a prototype in vertical pipes would be far more stable. Figure 5.12b depicts the robot during motion in vertical pipe with diameter ∅242 mm.

Pedipulators reconfiguration was analyzed in aspect of conformity with the mathematical model calculations. Figure 5.13a depicts comparison of transformation results from neutral pose to a pose for minimum size of vertical pipe with diameter ∅224 mm. We can observe that the absolute positioning error has significantly larger maximum value than in case of transformation to horizontal pipe. However, it occurs in the middle of transformation trajectory and can be neglected, since it is not a usable working range of motion. The same situation can be observed for transformation from pose for ∅224 mm pipe to ∅270 mm pipe (Fig. 5.13b) and for transformation from pose for ∅270 mm pipe to ∅224 mm pipe (Fig. 5.14a).

In Fig. 5.14b, results of maintaining attained pose for ∅270 mm pipe are depicted. The absolute positioning errors of V-REP joints positions, compared to the ones calculated from the mathematical model are negligible in this case.

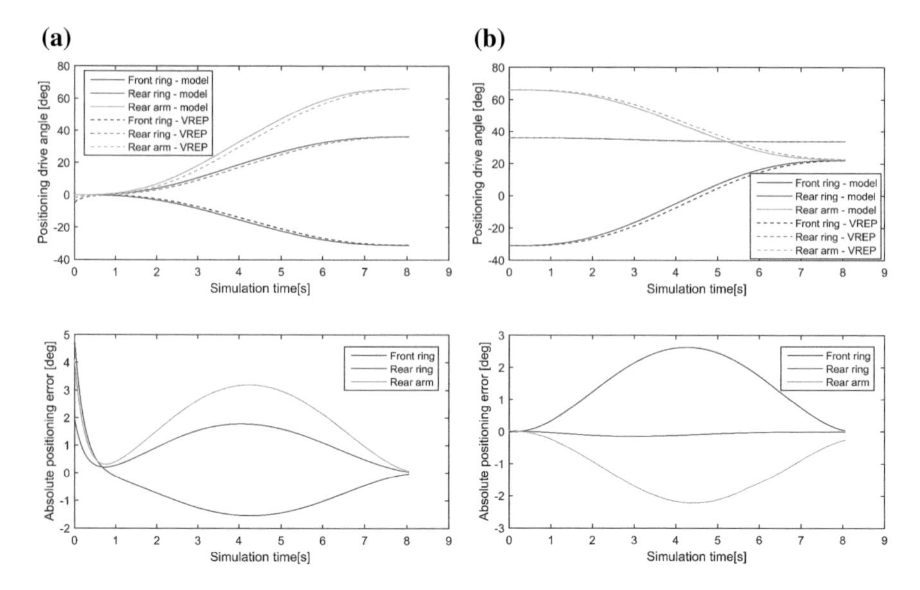

Fig. 5.13 V-REP and MATLAB simulations—transformation results comparison and absolute positioning errors: **a** from the neutral pose to a ∅22430 mm vertical pipe; **b** in a vertical pipe with diameter change from ∅224 to ∅270 mm

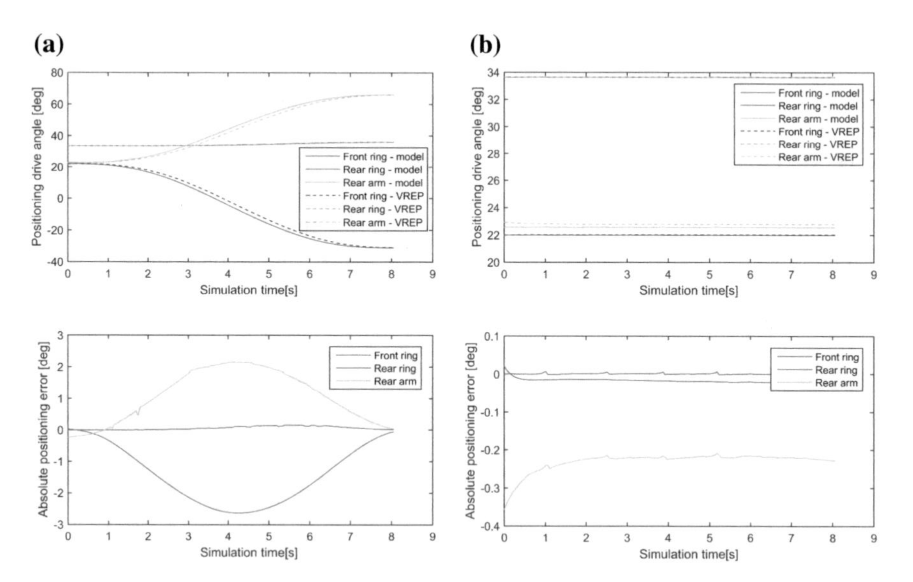

Fig. 5.14 V-REP and MATLAB simulations—transformation results comparison and absolute positioning errors: **a** from a ∅270 mm vertical pipe to a ∅224 mm vertical pipe; **b** maintaining constant pose in a ∅270 mm vertical pipe

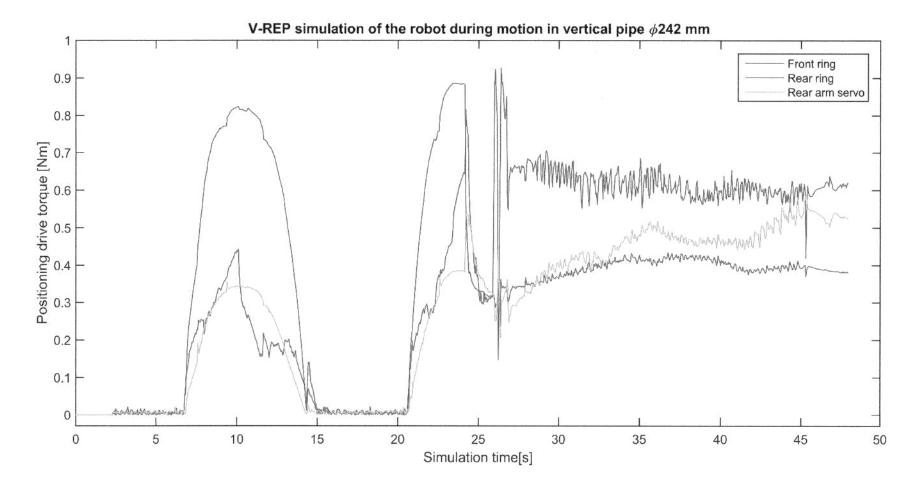

Fig. 5.15 Torques of the positioning servomotors during the V-REP simulation in a ∅242 mm vertical pipe

With the purpose of extension force control, torques of positioning servomotors were measured in the V-REP simulator. Torque values were used by the operator to automatically or manually adjust pedipulators' pose. In Fig. 5.15 we can observe the torque plot that depicts several phases of the simulation. At 3 seconds, transformation of the pedipulators from ∅224 to ∅242 mm pipe started and at 7 s, contact of tracks and pipe walls was initialized. At 10 s, we can observe that the robot attained desired pose and at 15 s, the contact was again released. At 21 s, the contact was reinitialized and at 25 s forward track motion started. Large alterations of torque in time between 25 and 30 s show centering of the robot in the middle of pipe and finally at 30 s, mean torque values stabilized with bounded fluctuation intervals. The fluctuations were caused by alternating contact between cylinders and pipe surface and would have significantly lower values in case of elastomeric track, used in the real robot prototype.

In Fig. 5.16 analogous torque plot is presented as in Fig. 5.15, however for the maximum vertical pipe diameter ∅270 mm. We may observe similar simulation phases and final stabilization when the robot moves in the center of pipe that occurs at 35 s.

An attempt was made to reconfigure the robot in a vertical pipe with smooth diameter change. In Fig. 5.17, we can see the reconfiguration phases. The initial pipe diameter of ∅224 mm was selected (Fig. 5.17a), whereas final diameter was set to ∅238 mm (Fig. 5.17b). The reconfiguration involved online manual control of the robot tracks extension on the basis of servomotors torque measurements, with usage of vertical pipe control blocks of the Simulink model depicted in Fig. 5.4.

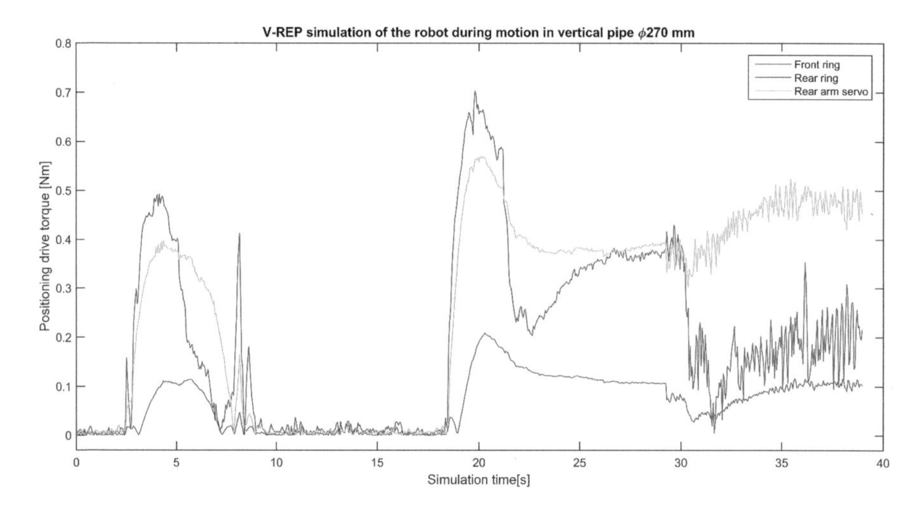

Fig. 5.16 Torques of the positioning servomotors during the V-REP simulation in a ∅270 mm vertical pipe

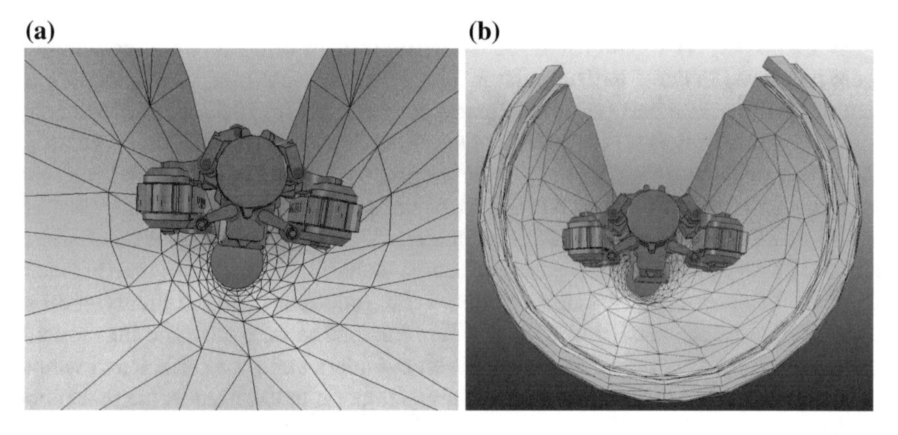

Fig. 5.17 V-REP model of the pipe inspection robot during reconfiguration in a vertical pipe: **a** initial diameter ∅224 mm; **b** final diameter ∅238 mm

Figure 5.18 shows a plot of positioning servomotors torques during simulation of the robot control in a vertical pipe with changing diameter. Between simulation time 1 and 9 s, the robot moves steadily in a section with diameter ∅224 mm. At 10 s we can observe the start of pipe diameter change. At 16.5 s the torque reaches critically low value, therefore manual change of track extension is applied. The same situation is visible at 25 and 33 s. Finally at 34 s the robot moves steadily in ∅242 mm pipe.

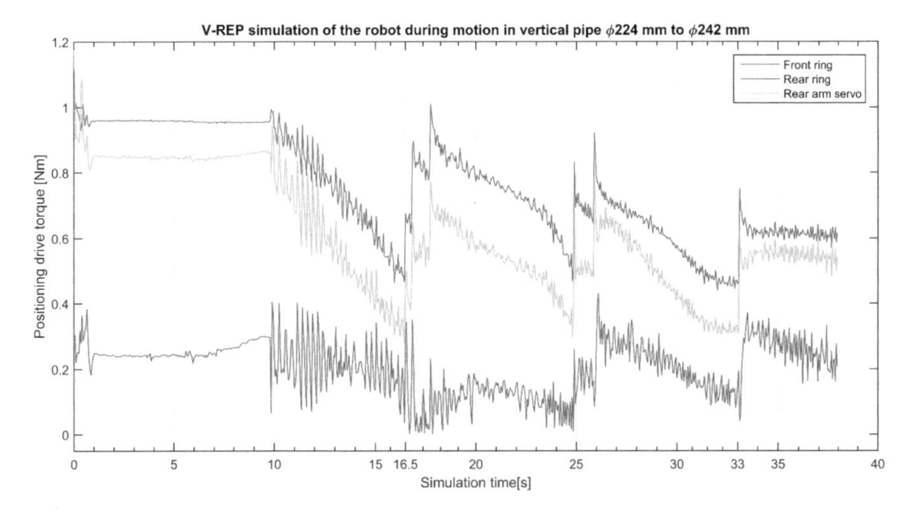

Fig. 5.18 Torque of the positioning servomotors during a V-REP simulation of the robot reconfiguration in a vertical pipe with smooth diameter change from ∅224 to ∅242 mm

5.6 Summary

Co-simulations of the robot control in V-REP and MATLAB environments were prepared to analyze and verify mathematical models of the robot, especially trajectory calculation for adaptation of pedipulators to different work environments. The simulations involved tests in horizontal pipes with different diameters and obstructions such as bends and reducers. Custom MATLAB/Simulink model was prepared that provides operator interaction with the simulation model and control hardware support.

Motion tests on even and rough surfaces were conducted that exhibited intrinsic properties of the pedipulator mechanism that requires specific transformation trajectories to goal poses, depending on orientation in space.

The simulations showed also that the robot is capable of transforming to inspect vertically oriented pipelines. Track extension investigation was performed, including error analysis and torque measurements. It was presented that reconfiguration of the robot in a vertical pipe with smooth diameter change is possible. This situation occurs very rarely in real life applications, since radical diameter changes by pipe reducers are more commonly encounted in pipelines. Nevertheless, the test shows that track extension can be manually or automatically adjusted, based on information about torques exerted by positioning servomotors, generating clamp force.

The simulations performed in V-REP and MATLAB provided valuable resources for control system of the robot prototype and served as indispensable validation tools for proper desing and tuning of the original pedipulator trajectory calculation algorithm.

References

1. Ivaldi S, Peters J, Padois V, Nori F. Tools for simulating humanoid robot dynamics: a survey based on user feedback. In: IEEE-RAS international conference on humanoid robots;2014. p. 842–849.
2. Yang C, Ye Z, Peter OO, Han J. Modeling and simulation of spatial 6-DOF parallel robots using Simulink and SimMechanics. In: 3rd international conference on computer science and information technology;2010. p. 444–448.
3. Nogueira L. Comparative analysis between gazebo and V-REP robotic simulators (Report). Universidade de Campinas;2014.
4. Ciszewski M, Mitka Ł, Buratowski T, Giergiel M. Modeling and simulation of a tracked mobile inspection robot in MATLAB and V-REP software. In: Postepy robotyki 1.195. Elektronika:Prace Naukowe PolitechnikiWarszawskiej;2016. p. 135–144.
5. Zhang X, Ivlev O. Simulation of interaction tasks for pneumatic soft robots using SimMechanics. In: 19th international workshop on robotics in Alpe-Adria-Danube region;2010. p. 149–154.
6. Spica R, Claudio G, Spindler F, Giordano PR. Interfacing Matlab/Simulink with V-REP for an easy development of sensor-based control algorithms for robotic platforms. In: 2014 IEEE international conference on robotics and automation workshop on MATLAB/Simulink for robotics education;2014.
7. Coppelia Robotics. V-REP virtual robot experimental platform. Accessed 14 Apr 2016. www.coppeliarobotics.com.
8. Khalilov J, Kutay AT. Interfacing Matlab/Simulink with V-REP for A controller design for quadrotor. Int. J. Eng. Res. Rev. 2015; 3(4):42–49.

Chapter 6
Control System Design and Implementation

6.1 Introduction

Control of mobile robots is a very broad scientific field that involves application of different techniques, software and hardware implementations. Mobile robots, in contrast to stationary industrial robots, may have morphologically dissimilar structure and can be driven by various actuators including electric motors, servomechanisms or internal combustion engines. The tasks of mobile robots that can explore environment with variable conditions may be also much more complicated than for industrial robots that usually have well defined workspace and task space [1]. Therefore, control of mobile robots needs to be divided into dissimilar task-oriented subgroups of procedures, software layers, subsystems and hardware-oriented architectures.

The mobile robot for pipe inspection that is presented in this work is a mechatronic device that requires application of two distinct control strategies. Its operation in workspaces such as flat surfaces or large diameter pipes involves usage of a control approach normally utilized for mobile robots, but transformation of pedipulators for environmental adaptation is based on mathematical apparatus applied for arm-type industrial robots. Motion control of the robot in pipelines is not a complicated task, assuming that there are no obstacles and geometry changes. For negotiation of elbows, reducers and sections filled with debris, more sophisticated track velocity control is required. In this chapter, a control system developed for the robot is described.

6.2 Control and Navigation of Mobile Robots

Mobile robots can be operated in different modes, according to their design and application. One type of control involves tele-operation by human. The human supervision can involve fully manual control of robot actuators, sensors and other equipment or may include semi-supervised execution of low-level control tasks by usage of

M. Ciszewski et al., *Modeling and Control of a Tracked Mobile Robot for Pipeline Inspection*,
Mechanisms and Machine Science 82, https://doi.org/10.1007/978-3-030-42715-3_6

predefined processes, subroutines and simple locomotion algorithms [2]. An autonomous mode of operation is the most demanding control strategy of mobile robots, when they can only rely on the information gathered by sensors to navigate through an unknown environment without direct operator supervision. In majority of cases, navigation is one of the most important aspect of mobile robots control. The knowledge about actual position and orientation of a mobile robot in space is usually essential for directing a mobile robot to its goal pose. Map creation is a significant part of robot navigation in an unstructured environment with obstacles. It helps in trajectory planning and augments exploration tasks by resources of useful information. Due to the fact that mobile robots have notably dissimilar structure, control approach should be applied according to specific group of robots. For instance, ground mobile robots, UAVs, underwater exploration ROVs or planetary rovers would require completely different algorithms and control system structure.

Motion control based on a kinematic model of a robot, called kinematic control may be difficult for nonholonomic systems such as the majority of ground mobile robots. The aim of a kinematic controller is to follow a trajectory, given its position or velocity data as a time function. It is convenient to divide the trajectory in motion segments of defined shapes such as straight lines or arcs with defined radius. Next, a smooth trajectory is precomputed for each trajectory segment between initial and final poses of a robot [3]. This control approach is an open-loop motion control, since the measured position and orientation are not used as feedback for position or velocity controller. Control laws for skid-steering differential drive robots are given in [2] for different motion scenarios. Exemplary application of this method for a 2-wheeled mobile robot is described in [4]. The limitations of this method involve difficulties of calculation of feasible trajectories, lack of adaptation to dynamic changes in the environment and jagged transitions between adjacent trajectory segments.

In contrast, a feedback controller provides information about real state of the robot, thus the path planning task can be reduced to setting intermediate positions that belong to the trajectory. By usage of feedback information about current pose of the robot, error information can be calculated [3]. An example of a closed-loop control of a 4-wheeled skid steering robot is given by [5]. The control approach is similar for a track drive robot.

Mobile robot navigation is a problem of guiding a robot to its goal position in a complex environment. Navigation tasks can be performed using two fundamental approaches: reactive navigation and map-based navigation. The first method is when a robot is reacting directly to environmental changes e.g. line-following, obstacle avoidance, light-intensity navigation. Commercial robots such as vacuum cleaners, lawn movers or swimming-pool cleaners use this type of navigation [6]. A good example of reactive navigation is a robot with two light-intensity sensors that is driven by light source to its goal pose. This method can be treated as a motion in a potential field towards the lowest (or highest) potential. Other type of reactive navigation is used by Bug algorithms that rely on moving in environment with obstacles to a predefined goal. In this case, the robot moves in a defined direction until it reaches an obstacle. These methods suffer from potentially non-optimal trajectories, since a map of the environment is not known, but reactive navigation is essential in dynamically changing environments. Map-based navigation is advantageous in complex environments and provides much more optimal trajectory planning. There exist multiple methods for map representation. The simplest one is an occupancy grid [6]. In this approach, a map is constructed as a set of cells that are marked as occupied by obstacles or unoccupied, through which the robot is free to plan a trajectory and move. The size of a grid cell is dependent on robot size and application. Map-based navigation with usage of an occupancy grid for a 2-wheeled mobile robot that is using distance transform for calculating the shortest path to a goal position is described in [7]. It is a simple and robust method for small maps but may suffer from high computational requirements for large datasets. An application concept of this type of navigation for a differential drive, 4-wheeled planetary rover is presented in [8]. Another algorithm for path planning with usage of an occupancy grid is called D* and is an extension of the A* algorithm. This method represents map cells with cost values of traversing each cell in horizontal and vertical direction. Obstacles are marked as infinite cost values, whereas e.g. rough terrain sections can have a finite cost value, depending on the used robot. The D* algorithm searches a path that minimizes total cost of achieving the goal position. The advantage of this algorithm is the incremental path replanning that is essential in dynamically changing environments. Usage of the D* algorithm for robot path planning is presented by [9]. Both previously described methods have major disadvantage when the goal position is changing, because the entire planning phase has to be repeated. To solve the problem, roadmap methods are used that base on building a network of obstacle free paths through the environment. The Voronoi diagram is an efficient method for creating a roadmap and path planning in an environment with changing goal position. Navigation on a roadmap is to choose the shortest path that leads to the vicinity of the goal position and then traverse to the goal [10]. The methods presented above are valid for an occupancy grid with sufficient cell size that gives the possibility to neglect nonholonomic constraints of a mobile robot and rely on motion controller to plan the trajectory between adjacent cells. Another approach, the RRT (Rapidly-exploring Random Tree) planner, takes into account the motion model of a vehicle. It computes discrete set of possible configurations that a nonholonomic vehicle can reach from its starting pose. From every computed pose the same procedure can be executed again. For an arbitrarily selected goal pose, it is possible to find the closest precomputed pose and by going backward to the starting pose, determine a sequence of steering angles and velocities needed to plan a trajectory. The RRT path planning approach for a 2-wheeled mobile robot is described by [11].

Map-based navigation and closed-loop control requires localization of the robot in the environment and on the available map. Map creation process is also required for some applications such environment exploration where ready to use maps are not accessible. Localization of a robot in the environment can be done by dead reckoning method based on its estimated speed, direction and time of travel with respect to previous estimate [6]. Odometry information, mainly obtained with usage of wheel (or track) motors rotation sensors and mathematical models of a robot provides location estimate of a robot in space, but it is prone to systematic errors due to imperfect wheel velocity, gyroscope calibration etc. Random errors such as slip of wheels or tracks, significant in case of differential drive vehicles and tracked robots also increase uncertainty of robot's pose. Practical application of odometric data for long trajectories and changeable surface conditions is impractical as a standalone method [10] but serves very well in indoor applications and flat surface operation of e.g. 2-wheeled robots [12]. For better estimation of robot pose with usage of the odometric information, Kalman filter is used that is a recursive algorithm. It calculates at each time step an optimal estimate of a robot pose, based on previous estimate and noisy odometric data. Since most robots are non-linear systems, the Extended Kalman Filter (EKF) method is used [6]. The EKF can integrate various sensory data for better estimation, including gyroscopic measurements, accelerometer readings, compass and GPS (Global Positioning System) measurements – when available.

To create a map of environment with obstacles, a robot can use odometric data and distance sensor measurements. The problem of concurrently determining position and creating a map is called in robotics Simultaneous Localization and Mapping (SLAM). In this problem, initial position of a robot or an estimate of the position is known and a map of an unknown environment is created during exploration phase [13]. The advantage of the SLAM technique over dead-reckoning is that the position estimate is improved by recognition of known obstacles. Since the map is stored, map-based navigation in an unknown terrain is possible. Indoor navigation with usage of a 2D-SLAM is described by [14, 15]. Currently, it is the most widely used mapping technique for mobile robots.

Navigation of mobile robots in pipelines is a different application problem, where some conventional methods are used, however this specific environment imposes usage of modified techniques. Latest research in the field of pipe inspection robots control has been made by some teams, featuring e.g. a wheeled mobile robot for inspection and pipe-mapping [16]. The robot is designed for movement in horizontal pipes and utilizes a forward facing fisheye camera. This setup allows to perform visual odometry and build pipe models with high resolution. Usage of a wide-angle camera with structured light and utilization of sparse reconstruction of straight segments and T-junctions (Fig. 6.1a) allowed to build a 3D rendered, shaded model of pipelines for localization of defects, as depicted in Fig. 6.1b.

Pipe navigation with usage of a 2D Laser Range Finder is described in [17]. The authors of this paper presented a method that relies on roll and pitch angles of the robot relative to horizontal plane and data from a 2D LIDAR, mounted on a rotational support. As an output, a 3D point cloud, reconstructed from the measurements is produced, as shown in Fig. 6.1c. This method provides estimation of pipe shape deformation and robot navigation in pipelines.

Another navigation concept, based on a vision system is presented in [18]. The authors showed that recognition of special features of pipelines such as elbows or branches is possible for usage as navigation landmarks. Their method focused on obtaining consistent shadow Images/images from a camera, equipped with uniform illumination. By analyzing the shadows, detection of passage direction, estimation of robot position and map reconstruction were possible to execute.

Utilization of SLAM techniques for navigation and map building is advantageous in pipes, with large number of characteristic features, however, in pipes where

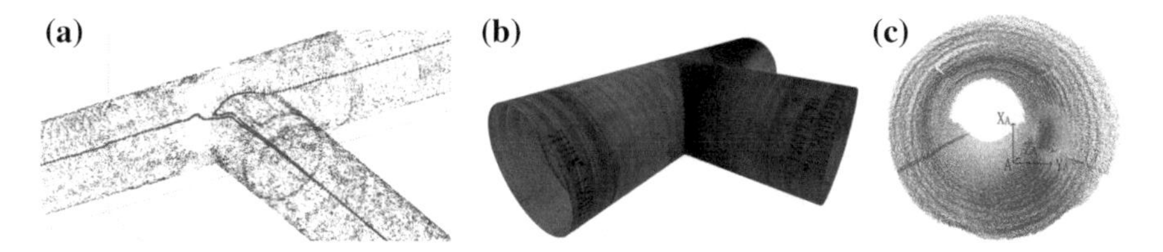

Fig. 6.1 Pipe navigation—mapping techniques: **a** visual odometry after sparse reconstruction [16]; **b** 3D render of a reconstructed pipe model [16]; **c** 3D point cloud of a pipe [17]

geometric changes are hardly noticeable, this technique is problematic due to accumulation of sensor errors as described in [19].

An autonomous pipe inspection robot that uses ultrasonic distance sensors and can connect wirelessly to the operator's control pendant was designed by Red Zone Robotics [20]. It is intended for operation in horizontal pipes and can navigate without a map, with usage of reactive control strategy. Its operation is dependent on the preselected work, chosen by the operator.

As we can see, navigation in pipelines requires alternative techniques to these used in e.g. ground mobile robots. Usage of Global Positioning System is not possible for localization, restricted space and insignificant changes in geometry decrease application possibilities of SLAM techniques or common trajectory planning algorithms. Yet, simple geometry and uncomplicated structure of the environment reduces the need for precise localization that can be limited to odometric measurements from wheel or track drive sprocket revolutions to locate the robot along pipe length.

6.3 Control of Arm-Type Robots

Motion control of manipulators and other arm-type robots is a challenging task due to the complexity of dynamic models and uncertainties that arise in physical systems. Dynamic description of a robot is difficult mainly due to nonlinearity and motion coupling between consecutive links. Uncertainties, may arise from insufficient knowledge about dynamic parameters of robot structure or can result from joint and link flexibility, backlash, actuator dynamic properties, friction, sensor measurement errors and environmental disturbances [10]. Classical approach for control of manipulators is to use dynamic model of a robot. The model is usually prepared with usage of Lagrange's equations of 2nd kind with the assumption that the robot is fully actuated; it means that there is an independent control input for each degree of freedom. To formulate dynamic equations of motion, link masses and inertia matrices have to be specified, next the equations can be derived based on the robot's structure [10]. The main control objectives that may arise from operation scenario of a manipulator include: trajectory tracking and regulation. Trajectory tracking is aimed at following of a time-varying joint reference trajectory that is defined within the robot workspace. In this case limits of velocity, acceleration and torque of drives should not be violated. The objective of regulation is to provide a point-to-point control between fixed configurations in the joint space. Thus, joint variables have to be set and maintained at desired positions, independently of disturbances and initial positions. Application of the control system is task-dependent and structure-dependent. Robot joint space control for tasks that demand high precision during entire motion require trajectory tracking control (e.g. welding, gluing, machining), whereas for tasks in which the manipulator is set to precisely attain initial and goal poses (pick and place), regulation control may be used.

In motion control of manipulators, individual tasks such as pick and place operations form a larger, higher-level sequences. Whilst simple moves are executed using joint space control, more complex sequences that take into consideration trajectories, necessary to avoid obstacles, require operational space control (task space control).

The practical design of joint space control is to formulate a feedback controller such that joint coordinates track the desired motion as closely as possible. The general scheme of joint space control is presented in Fig. 6.2. Initially, end-effector trajectory in e.g. Cartesian coordinates x_d is converted by inverse kinematics model to desired joint coordinates q_d that are input to the controller. The controller, with usage of the dynamic model of the robot outputs a vector τ of torques and forces to the manipulator, where actual joint coordinates q are measured and fed back to the controller. This control strategy is valid when manipulator trajectory can be preplanned and it is not changing during motion. The joint space control methods include, among others: PD control, PID control, inverse dynamic control, passivity-based control [10].

In case when manipulator workspace is not well defined or it is dynamically changing during motion, operation space control strategy is used. In this approach, end-effector trajectory can be modified online to avoid suddenly appearing obstacles or altered according to other sensor inputs. This strategy is derived directly on the dynamic model of the robot, represented in the operational space. Based on the desired end-effector trajectory x_d, the controller uses operational space dynamic model to calculate command forces f_c that are fed to manipulator actuators as shown in Fig. 6.3. Actual end-effector position x is returned to the controller.

The advantages of task space controllers over joint space controllers include direct task error minimization by feedback loop and that direct calculation of inverse kinematics is not required, because velocity kinematics are included in the model [10].

Different control techniques are used depending on robot application. Independent joint control, in which particular joints are controlled individually is computationally effective and scalable between different actuator designs. PID-type control can

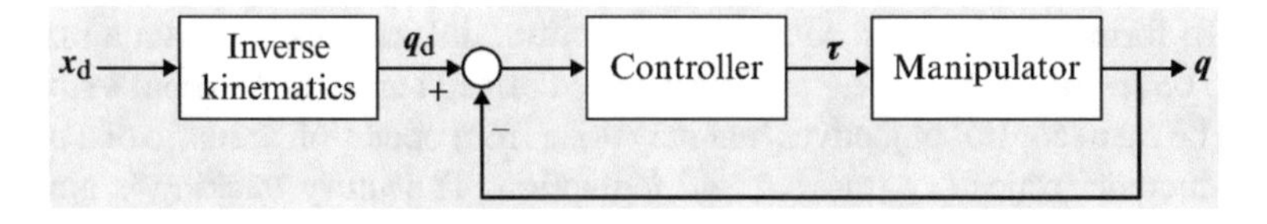

Fig. 6.2 Joint space control—general scheme

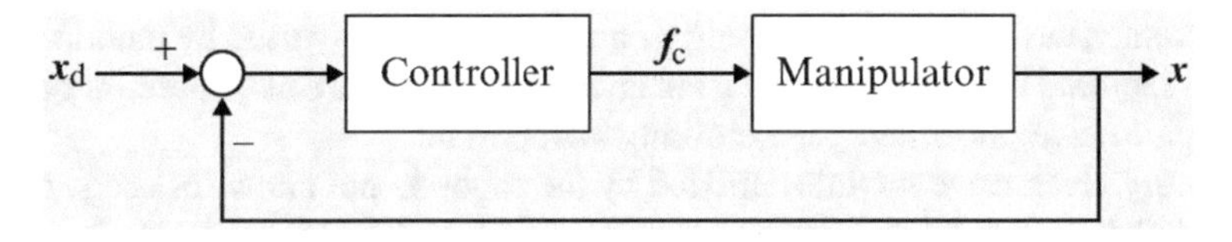

Fig. 6.3 Task space control—general scheme

solve the regulation problem and has advantage when model structure and model parameters are not available. For complex tasks that require different velocities and accelerations for particular subtasks such as in painting process, where manipulator approach should be realized faster than painting, it is advantageous to use tracking control. It uses trajectory and its derivatives to describe desired velocity and acceleration of control outputs. Computed-torque control can be regarded as applying feedback linearization to nonlinear systems. When dynamical model of a robot is known, they can lead to convergence of tracking error to zero, but require high computational power. An adaptive control approach allows to use time-varying controller parameters to react to changing environment and structural and non-structural uncertainties in the robotic system. Optimal controller design is possible when an accurate model of the manipulator is available, but can lead to instability if modeling errors are encountered. Robust controllers are used for systems with unmodeled dynamics. The learning control method involves storage of data from previous cycles, thus it features iterative improvement of performance for repetitive tasks.

Selection of a control technique is highly dependent on task specification, desired kinematic and dynamic performance, available actuators and controller hardware.

6.4 General Structure of the Pipe Inspection Robot Control System

The tracked mobile robot for pipe inspection, described in this work is a hybrid mechatronic device that does not simply belong to mobile robots in terms of control strategy that should be applied for its proper operation.

Firstly, analogously to other pipe inspection robots, its main operation environment requires usage of a cable for control, due to the fact that wireless communication in pipelines may be generally problematic. Some research was conducted on wireless communication inside of pipelines according to [21], however quality of data transfer is dependent on pipe material, diameter, network layout and ground properties. It is possible to place antennas inside of pipe to enhance communication quality, but it is challenging for longer distances [22] and current technology that relies on radio frequencies does not provide online video transmission from a CCTV camera that is the main inspection tool of the discussed robot.

Secondly, navigation and mapping techniques, commonly used by mobile robots are not required in simple pipelines, but alternative methods can be used for complicated pipe networks, where pipe profiling is important.

Thirdly, track drive modules utilized by the robot do not allow for usage of precise odometric data when moving inside of pipelines. Therefore, the desired operation scheme implies application of tele-operation for robot motion control inside of pipelines. The manual control of track drives by the operator is possible, because video feedback is supplied from the CCTV camera.

Finally, the most challenging control task for the pipe inspection robot is the pedipulators reconfiguration to work environment that is rather related to control techniques of arm-type robots than mobile robots. This operation can be performed on the basis of previously described original algorithm, presented in Sect. 4.4. The control method, utilized for the reconfiguration is the joint space control, where final and initial poses are given and joint space trajectories are calculated. Actuators of the pedipulators mechanisms possess PID controllers, thus it is satisfactory to supply data with positions at consecutive trajectory steps. In this case, application of task space control would require encoders mounted on each rotary joint of the pedipulators and it is not included in this design.

All these aspects, described above lead to development of a custom, original control approach, specific for this patented robot structure.

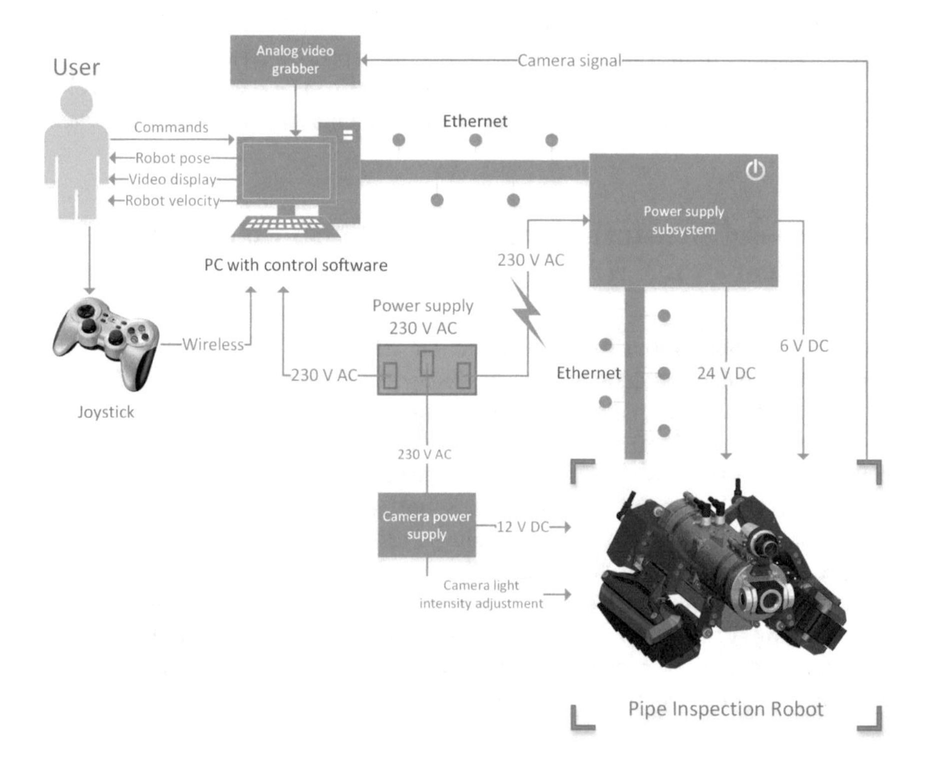

Fig. 6.4 Control system architecture of the pipe inspection robot

Hardware overview of the robot control system architecture is presented in Fig. 6.4. The robot is supplied by 24 V DC and 6 V DC voltages and communication to the power supply subsystem is realized via an Ethernet cable. Additionally, the camera is supplied by 12 V DC voltage and has manually controlled light intensity adjustment. The power supply subsystem is dedicated for 230 V AC mains voltage. Control signal of the robot is routed to a PC with software and analog camera signal is fed to a video grabber, also connected to the PC. User duties for controlling the robot include among others: setting desired pedipulators pose for particular pipe size, supervision of robot actual drive positions and track drive velocities. Track drives and pedipulators reconfiguration can be controlled by a joystick, connected wirelessly to the PC control software. Manual operation of the robot is realized as a supervised tele-operation with video feedback from the CCTV camera. Transformation of the robot pedipulators is realized automatically, based on the trajectories, generated with usage of the mathematical model.

The control system consists of components specially designed for the pipe inspection robot, including electronic control board, power supply subsystem, low-level control software, installed on robot's microcontroller and high-level software implemented on the PC computer.

6.5 Electronic Control Board

Control of the robot actuators and inspection equipment requires a dedicated electronic system. Due to the fact that the robot is designed to operate on a tether cable, connected to external power supply and a computer with control software, it is necessary to utilize an on-board controller, located inside of the robot body. The majority of pipe inspection robots, presented in the literature review feature significantly lower number of actuators and do not require a highly-specialized control unit, but can be designed based on widely available solutions for control of motors and inspection equipment. The complex motion unit of the robot that includes two different drive types, requires a dedicated Printed Circuit Board (PCB) with master level microcontroller and low-level chips that operate particular motors. Requirements for the electronics control board are listed in Fig. 6.5. It was decided to utilize Ethernet communication with the PC computer. The board provides individual velocity control of DC motors, located in the track drive modules by Pulse Width Modulation (PWM) method, utilizing length of duty cycle to control the velocity of tracks. Additionally, the board hosts 6 PWM channels for position control of servomotors. This control is realized by setting appropriate width of signal that corresponds to particular motor position. In case of the Hitec servomotors, utilized in this design, pulse width should be adjusted between 750 and 2250 μs, with 1500 μs as middle position. The pulse refresh frequency is 50 Hz. The PCB is additionally equipped with an IMU that features 3-axis accelerometer and 3-axis gyroscope. The IMU has an integrated EKF (Extended Kalman Filter) for pitch and roll angle estimation and FIR (Finite Impulse Response) filter for noise reduction.

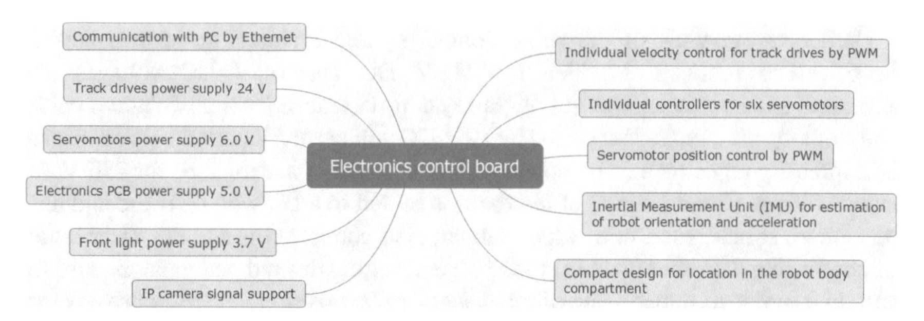

Fig. 6.5 Design requirements of the robot on-board controller

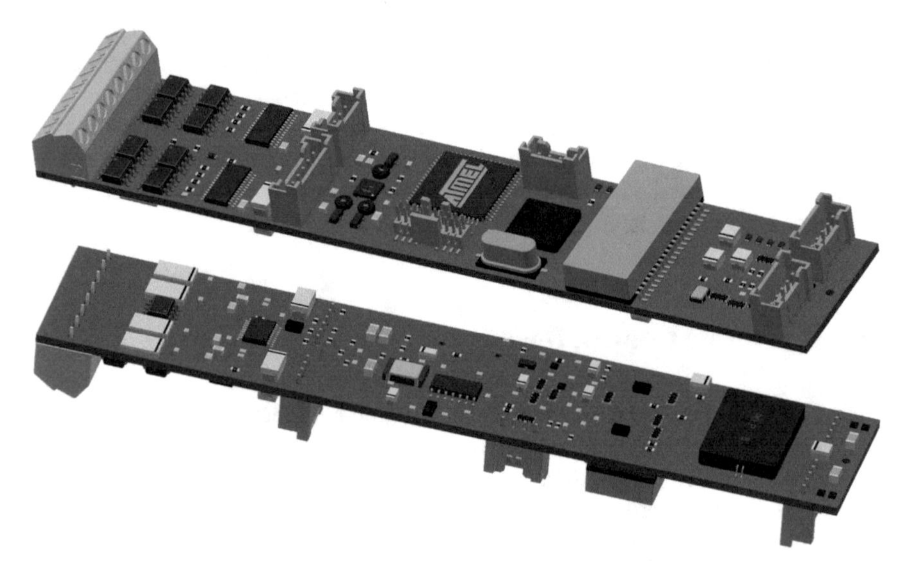

Fig. 6.6 The electronic control board located in the robot body—a 3D model of the PCB

The electronic control board was designed and modeled to verify conformity with mechanical design of the robot assembly. The 3D model of the board is presented in Fig. 6.6. Architecture of the PCB was planned, based on the smallest available components that possess required technical parameters for robot control tasks.

Dimensions of the PCB with mounted components are $145 \times 28 \times 14$ mm. It is fitted in the bottom compartment of the robot body. External cable connections with the power supply and communication unit are realized by two waterproof connectors located in the back of robot body.

6.6 Software Implementation

Software implementation for the robot control system consists of two applications. The first one is written in C language and it is dedicated for Atmel ATxmega128—AVR architecture microcontroller. It operates at frequency 32 MHz and it is located on the robot PCB board. The main tasks of the microcontroller include Ethernet communication handling, generation of two types of PWM signals, according to user inputs, transmission of data gathered from Inertial Measurement Unit to control software on a PC computer. Ethernet communication is realized with usage of the TCP IP protocol and Wiznet Data Handler driver. The robot PCB board has a statically assigned IP address. User commands are transferred from the PC computer to the PCB, with usage of packets containing motion commands, whereas IMU measurements and camera signal are fed back to the computer. The PWM generation is divided into two types that run in Single Slope PWM Waveform Generation mode. The first type of PWM is dedicated for velocity control of track drives, where duty cycle length is varied from 0 to 100% and corresponds to zero and maximum forward velocity. Backward velocity is controlled with PWM of the same type and an H-bridge, set to reverse rotation state. The second type of PWM is used in pedipulators servomotors by setting appropriate pulse width, corresponding to desired actuator positions. The PID controllers, integrated in the servomotors assure the fastest possible motion to the desired position.

The second application for controlling robot operation is dedicated for the PC computer. It was designed in C++ language with usage of Microsoft Visual Studio. The main functionalities of the software include: communication with PCB control board by Ethernet, setting servomotors positions, setting track drives velocities, reading IMU data, displaying plots with accelerations and angular velocities, displaying CCTV inspection camera video, handling of multiple user input methods. The application was developed according to hardware requirements of the robot control system, specified by the author of this book. With usage of the application, it is possible to operate the robot with a joystick by manually setting track velocities and servomotor positions. It is also possible to read CSV files with saved pedipulators transformation trajectories [23]. At the stage of application development, trajectories for control of the robot arms were saved in manual teach mode that consisted of trial and error procedure to attain desired pedipulators poses. The control of servomotors was realized in pairs, so that track drives pose was set symmetrically for each side of the robot. This approach simplified training algorithms but in case of unequal backlash in pedipulators joints and positioning drives, it may lead to difficulties in setting appropriate poses of the track drive modules. The main window of the application is shown in Fig. 6.7. It consists of four sections. The first one, in the upper left side of the window consists of track drives velocity displays and servomotors positions displays. In the lower left side of the window, the main command buttons are located. Manual and automatic reconfiguration modes are available. The automatic reconfiguration of pedipulators consist of predefined poses, assigned for particular pipe sizes and additionally, "Play recorded path" option is available that allows to replay any trajectory

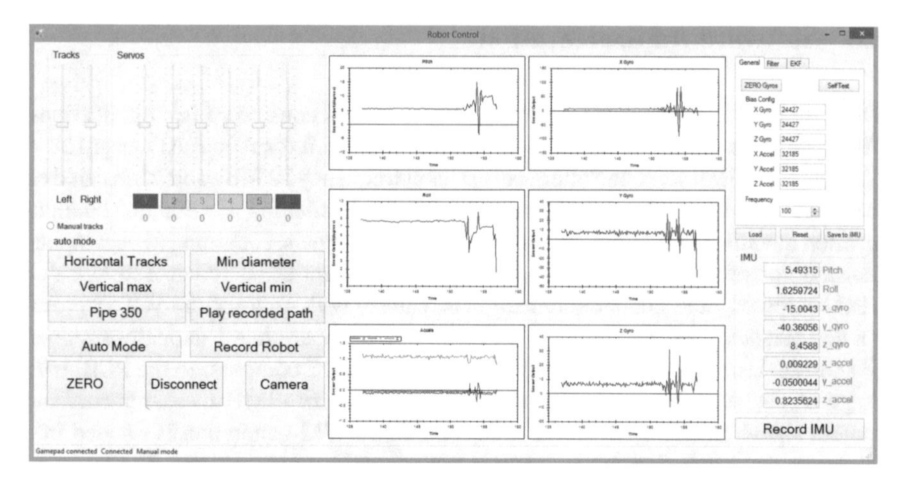

Fig. 6.7 The robot control software—main window [23]

of pedipulators, saved in a .CSV file. The "Record Robot" option allows logging of servomotors positions for future analyses and provides replay functionality. To comply with the transformation calculation algorithm, developed by the author of this book, it is necessary to use "Play recorded path" function with trajectory exported from mathematical model developed in MATLAB. The lower left buttons are used to reset servomotors positions to neutral "Zero" pose, manage connections and enable CCTV camera video display.

The middle section consists of six plots. Data, visualized by the plots is collected by the IMU sensor. These plots consist of Roll and Pitch angles of the robot, accelerations, and gyroscope outputs with respect to the X, Y and Z axes. On the right side of the window, the fourth panel is used for displaying of numerical values from the IMU, calibration and recording of measurements.

With usage of the presented control software, it is possible to reconfigure the pedipulators to adapt to particular work environment, perform pipe inspection, log robot motion parameters and replay recorded data.

6.7 Summary

The control system of the pipe inspection robot, presented in this chapter is based on several concepts, related to control of mobile robots and arm-type robots. Literature research of both scientific fields is followed by exemplary applications and relations to the designed control system. Structure of the control system is discussed in the aspect of communication, power supply, user interface and connections between different elements. Design of a dedicated electronic control board is outlined, including architecture, technical parameters and functionalities. Finally, software implementation is discussed on the level of on-board controller and high-level PC computer application with user graphical interface. Execution of the pedipulators trajectories is provided by the designed control system to transform the robot to different work environments.

References

1. Egerstedt M. Control of autonomous mobile robots. In: Handbook of networked and embedded control systems; 2005. p. 767–778.
2. Cook G. Mobile robots navigation, control and remote sensing. New Jersey: Wiley; 2011. (Chapter 6. Control system design and implementation 95)
3. Rol S, Nourbakhsh IR. Introduction to autonomous mobile robots. Cambridge, USA: MIT Press; 2011.
4. Klancar G, Matko D, Blazic S. Mobile robot control on a reference path. In: 13th mediterranean conference on control and automation; 2005, p. 1343–1348. (Chapter 6. Control system design and implementation 96).
5. Kozlowski K, Pazderski D. Modeling and control of a 4-wheel skid-steering mobile robot. Int J Appl Math Comput Sci. 2004;14(4):477–96.
6. Corke P. Robotics, Vision and control: fundamental algorithms in MATLAB, vol. 3. Springer Science & Business Media; 2011.
7. Buratowski T. Mobile robots-selected issues. 1st ed. Krakow: AGH University of Science and Technology Press; 2013.
8. Ciszewski M, Buratowski T, Uhl T, Giergiel M, Seweryn K, Teper W, Zwierzynski AJ. Ultralight mobile drilling system-design and analyses of a robotic platform intended for terrestrial and space applications. In: Robot motion and control (RoMoCo). 2015. p. 84–90.
9. Francis SLX, Anavatti SG, Garratt M. Dynamic model of autonomous ground vehicle for the path planning module. In: ICARA 2011—Proceedings of the 5th international conference on automation, robotics and applications; 2011. p. 73–77.
10. Siciliano B, Khatib O. Springer Handbook of Robotics, 1st ed. Berlin: Springer; 2008.
11. Palmieri L, Koenig S, Arras KO. RRT-based nonholonomic motion planning using any-angle path biasing. In: Proceedings of IEEE international conference on robotics and automation 2016-June; 2016. p. 2775–2781.
12. Buratowski T, Dabrowski B, Uhl T, Banaszkiewicz M. The precise odometry navigation for the group of robots. Schedae Inform. 2010;19:99–111.
13. Corke P. Robotics Toolbox. 2016. http://petercorke.com/Robotics_Toolbox.html. Accessed 10 Feb 2016.
14. Buratowski T, Ciszewski M, Giergiel M, Kudriashov A, Mitka Ł. Robot z laserowym czujnikiem odległości do budowy map 2D. Program 55. sympozjonu "Modelowanie w Mechanice", 2016.
15. Ciszewski M, Buratowski T, Giergiel M, Kudriashov A, Seweryn K, Teper W, Zwierzynski A, Uhl T. Inspection robot scaner system based on a LiDAR mapping solution in study of problems in modern science: new technologies in engineering, advanced management, efciency of social institutions. Khmelnytsky National University; 2015.
16. Hansen P, Alismail H, Rer P, Browning B. Visual mapping for natural gas pipe inspection. Int J Robot Res. 2015;34(4–5):532–58.
17. Hu Y, Song Z, Zhu J. Estimating the posture of pipeline inspection robot with a 2D Laser Range Finder. In: 2012 IEEE international conference on multisensor fusion and integration for intelligent systems (MFI); 2012. p. 401–406.
18. Lee J-S, Roh S, Kim DW, Moon H, Choi HR. In-pipe robot navigation based on the landmark recognition system using shadow images. In: 2009 IEEE international conference on robotics and automation; 2009. p. 1857-1862.

19. Yatim NM, Shauri RLA, Buniyamin N. Automated mapping for underground pipelines: An overview. In: 2nd international conference on electrical, electronics and system engineering (ICEESE); 2014. p. 77–82.
20. RedZone. Solo Unmanned inspection robot. http://www.redzone.com/products/solo-robots/. Accessed 21 Oct 2012.
21. Wu D, Ogai H, Yeh Y, Hirai K, Abe T, Sato G. Pipe inspection robot using a wireless communication system. Artif Life Robot. 2009;14(2):154–9.
22. Nagaya K, Yoshino T, Katayama M, Murakami I, Ando Y. Wireless piping inspection vehicle using magnetic adsorption force. IEEE/ASME Trans Mechatron. 2012;17(3):472–9.
23. Wacławski M. Control of arms of a mobile inspection robot. MA thesis, AGH University of Science and Technology; 2013.

Chapter 7
Prototype of the Pipe Inspection Robot

The mechanical design, mathematical modeling, simulations and development of electronics and control system of the robot led to creation of a complete documentation, required for development of a prototype. The prototype building process involved simultaneous construction of the mechanical subsystem, electronic control board prototype, development of low and high-level control software, communication system and power supply unit. Numerous experiments that allowed to verify operation of the prototype are described in this chapter. Finally, the experience acquired during these tests permitted to propose improvements, indispensable during creation of an enhanced version of the pipe inspection robotic system.

7.1 Development and Construction of the Prototype

The prototype of the robot was built, based on the documentation and 3D models prepared at the design stage. Application of the Autodesk Inventor 3D modeling software enabled to use Computer Aided Manufacturing (CAM) for generation of sequences for CNC machining equipment. Due to the shape complexity of the designed parts, caused by strict construction requirements, it was necessary to utilize wide variety of manufacturing techniques. The majority of parts were manufactured with usage of CNC machines that involved milling, turning and CNC Electro-Discharge Machining (EDM). The wire-cut EDM was necessary to produce internal meshing gear transmission components for coaxial rings, located in the robot body. This method was selected due to small dimensions and high precision requirements of these parts, machined in a unitary production scale. Some parts with less complicated shapes were manufactured with usage of manually operated milling and turning machines. Production of seals was performed with application of laser cutting process.

By reason of the fact that most of the robot components were manufactured from aluminum, it was necessary to apply additional surface protection to enable

M. Ciszewski et al., *Modeling and Control of a Tracked Mobile Robot for Pipeline Inspection*, Mechanisms and Machine Science 82, https://doi.org/10.1007/978-3-030-42715-3_7

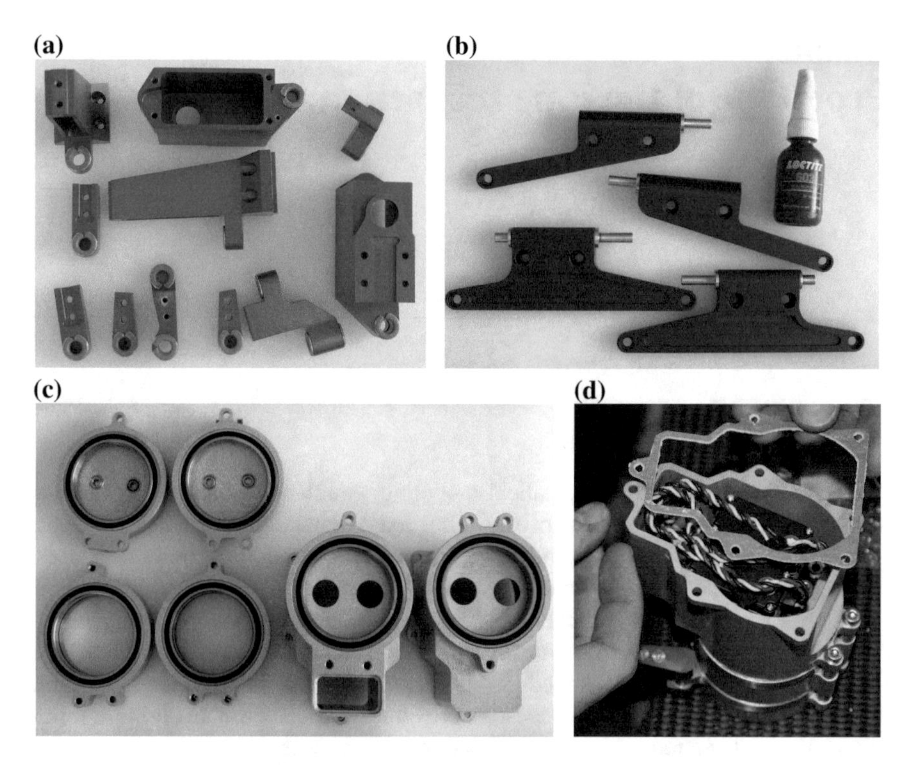

Fig. 7.1 Aluminum parts of the robot—anodized: **a** arm elements; **b** track mounts; **c** body elements; **d** half of the robot body assembly with a rubber seal

long-term operation of the prototype in water or other liquid environments. All aluminum components were polished and then anodized. To improve visual aspects of the prototype, colors of the components for the anodization process were chosen to suit the 3D model and distinguish pedipulators structure from the rest of the robot assembly.

The aluminum components after manufacturing, anodization and assembly of bearings are shown in Fig. 7.1. Arm elements with flanged DU-bushings are shown in Fig. 7.1a. The bushings are used to minimize friction between rotating rings and pins, made of stainless steel. Red-colored elements (Fig. 7.1b) are used for mounting of the track drive modules to the robot arms that are anodized in blue color. Yellow-colored components that form the robot body, along with assembled bearings for fixing rotating rings are shown in Fig. 7.1c. The robot body is divided into two half-assemblies that are connected with screws and sealed with a face rubber seal (Fig. 7.1d). Assembly of the ring drives to the robot body required calibration of the servomotors neutral, minimum and maximum positions with rotating rings angles (Fig. 7.2a).

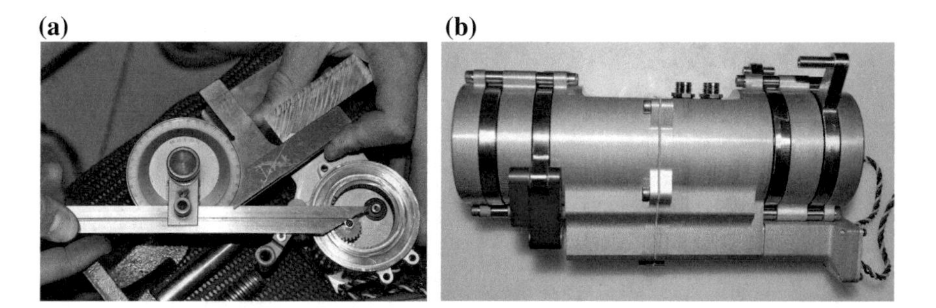

Fig. 7.2 Assembly process of the robot body: **a** calibration of the servomotors with the ring drives gear transmissions; **b** the assembled robot body

The calibration was conducted to match the mathematical model, used for control of the robot. The assembled robot body is presented in Fig. 7.2b. Finally, the assembly process was successfully performed and mechanical components of the robot prototype were integrated with actuators.

7.2 Prototype Comparison with the CAD Model

Assembly of the prototype was followed by comparison of the robot with the CAD model to eliminate all possible discrepancies between a virtual concept and a physical robot. The main objective of the verification was to check ranges of positioning drives by manually setting desired pedipulators poses, assumed in the mathematical model. Comparisons of the CAD models and the prototype with manually set pedipulators poses for motion in horizontal pipes with circular cross-section are shown in Fig. 7.3.

Figure 7.4 depicts the CAD model and the prototype in a pose for operation in horizontal, rectangular pipes and horizontal surfaces, whereas Fig. 7.5 depicts the comparison in the case of vertical pipes.

It can be observed that all predefined poses, presented in the 3D CAD models were manually set on the prototype, when all motors were turned off. This test showed that the prototype is ready for implementation of the control system and application of pedipulators transformation trajectories, obtained with usage of the original algorithm.

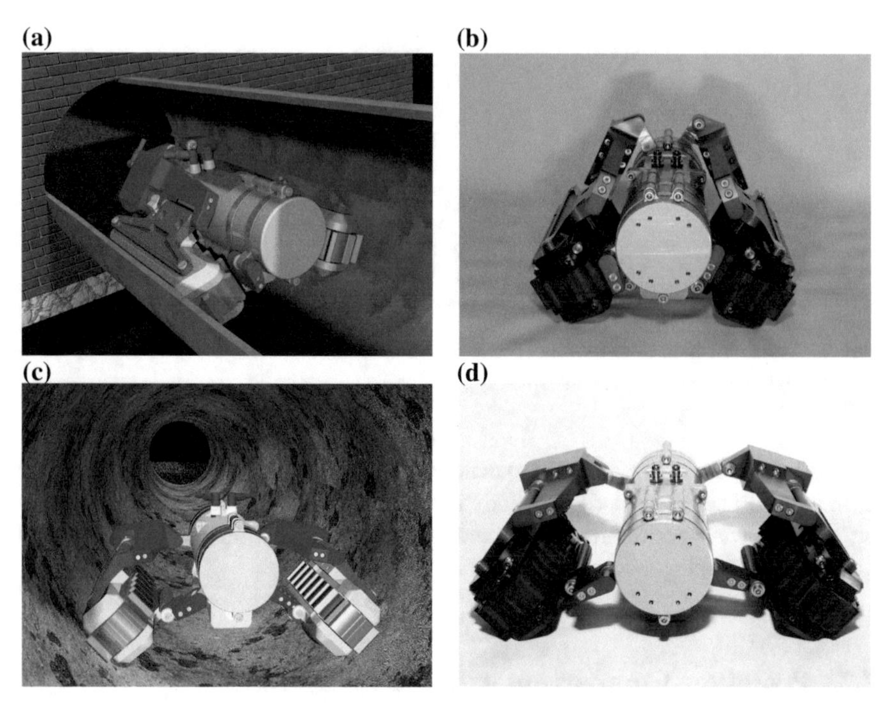

Fig. 7.3 Comparison of the CAD model and the prototype—pedipulators poses for horizontal pipes: Ø210 mm **a** CAD model, **b** prototype; Ø300 mm **c** CAD model, **d** prototype

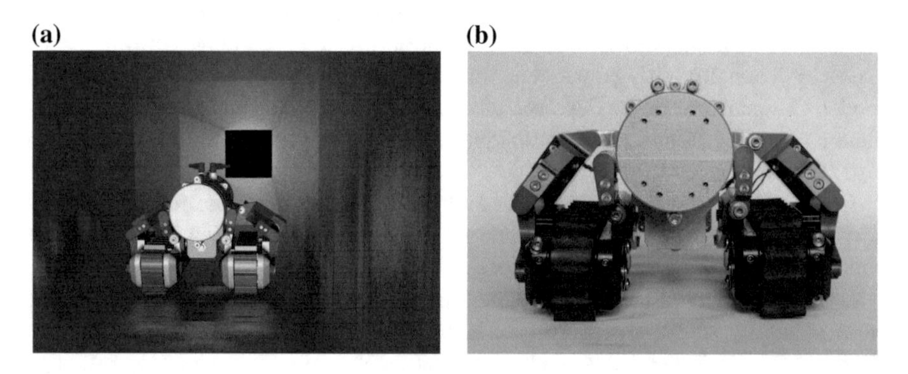

Fig. 7.4 Comparison of the CAD model and the prototype—pedipulators poses for horizontal surfaces: **a** the CAD model; **b** the prototype

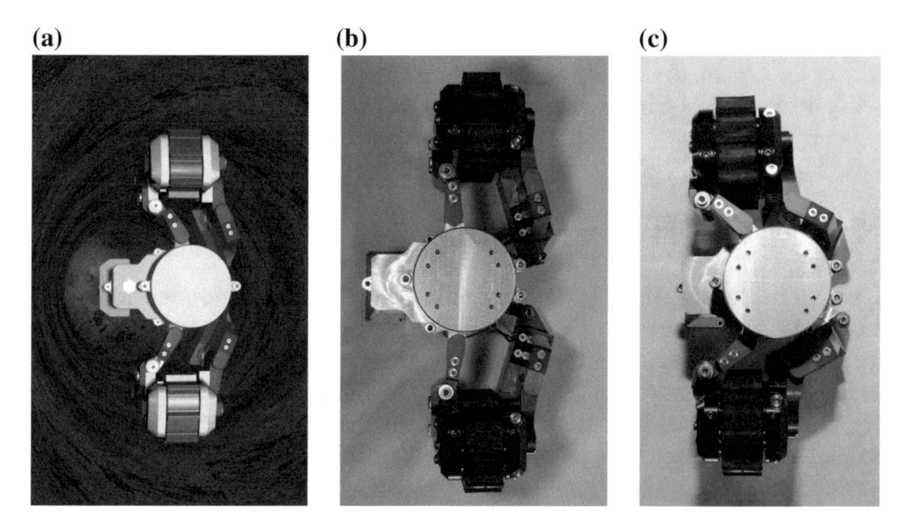

Fig. 7.5 Comparison of the CAD model and the prototype—pedipulators poses for vertical pipes: **a** Ø250 mm—the CAD model; **b** Ø270 mm—the prototype; **c** Ø224 mm—the prototype

7.3 Electronic Components and Control System of the Robot

Control system of the robot prototype was integrated, according to the requirements, specified at the design stage. The PCB control board was mounted inside of the robot body with all necessary connections and test cables were attached to two waterproof connectors, located in the back of the electronics compartment. Power supply for the prototype was placed in an ABS plastic housing and adapted to 230 V AC mains voltage. In the power supply box, two voltage transformers were located. The first one, rated at 200 W generates 24 V DC voltage for track drives and the second one, rated also at 200 W, supplies servomotors, PCB control board and supplementary light with 6 V DC voltage. One 4-wire cable is used for power supply, whereas, the second, 12-wire cable is utilized for Ethernet communication. Analog camera cable is routed separately. Control system of the prototype includes also a computer with Windows operating system, LAN network adapter and a wireless joystick. Communication and supervised control of the robot is realized with usage of the software, described in Sect. 6.6.

7.4 Laboratory Tests of the Prototype

Building process of the prototype was completed including: mechanical structure, actuators, electronic control unit, communication subsystem, power supply, inspection equipment and software with implementation of robot motion control. The building process was followed by numerous experimental procedures to assess functionality of the design, check conformity of the prototype with stated requirements and verify mathematical models with control system implementation.

7.4.1 Pedipulator Mechanism Reconfiguration

The first tests involved individual verification of eight robot actuators including track drives and servomechanisms. The tests were followed by laboratory verification of robot pedipulators transformation. To freely attain all desired poses of the pedipulators, it was necessary to place the prototype on a dedicated support as depicted in Fig. 7.6. In order to perform these tests, results of the mathematical modeling of the pedipulators were used. The original trajectory calculation algorithm was applied to generate joint trajectories for all servomechanisms, located in the robot arms. The trajectories were previously validated by simulations of the mathematical model and further converted to CSV files with time series of joint coordinates that can be interpreted by the software. In Fig. 7.6, attained poses of the robot prototype that are used during motion in horizontal pipes and flat surface are presented. Figure 7.6a depicts the prototype in the neutral pose that is set after startup of the on-board microcontroller. Figure 7.6b shows the prototype after realization of the trajectory, necessary to attain a pose for motion on horizontal surfaces or in rectangular pipes and ducts. It should be noted that this pose is one of the most difficult to attain from the neutral pose, presented in Fig. 7.6a, since it requires an intermittent pose for proper generation of the trajectories by the pedipulator motion control algorithm. Figure 7.6c displays the prototype in the most compact pose that will be utilized for inspection of horizontal pipes with diameter Ø210 mm. The pedipulators, transformed to operate in horizontal pipes of greater diameters e.g. Ø315 mm, are depicted in Fig. 7.6d.

Reconfiguration of the pedipulators to realize prototype motion in vertical pipes is a novelty, since there do not exist similar inspection robots that can be as versatile in such wide variety of environments. The generation of trajectories was performed in an analogous way as for horizontal surfaces and pipes. Figure 7.7 depicts the robot after completion of the trajectory from neutral pose to the pose used for motion in vertical pipes with diameter Ø224 mm. This is the most compact pose of the robot for vertical operation. The maximum extension of pedipulators for vertical motion is depicted in Fig. 7.7b, allowing to move in Ø270 mm pipes. It should be noted that motion in vertical pipes requires considerably high clamp force generated by extension of tracks. The theoretical pipe diameters do not take into consideration mechanism backlash and track tread deformation during clamping to pipe walls. Thus, the actual maximum pipe diameter would have slightly smaller value. In order to attain a stable position of the robot prototype in a vertical pipe before execution of inspection tasks, it is necessary to reconfigure the pedipulators to the minimum pose and gradually extend track drives to obtain necessary clamp force.

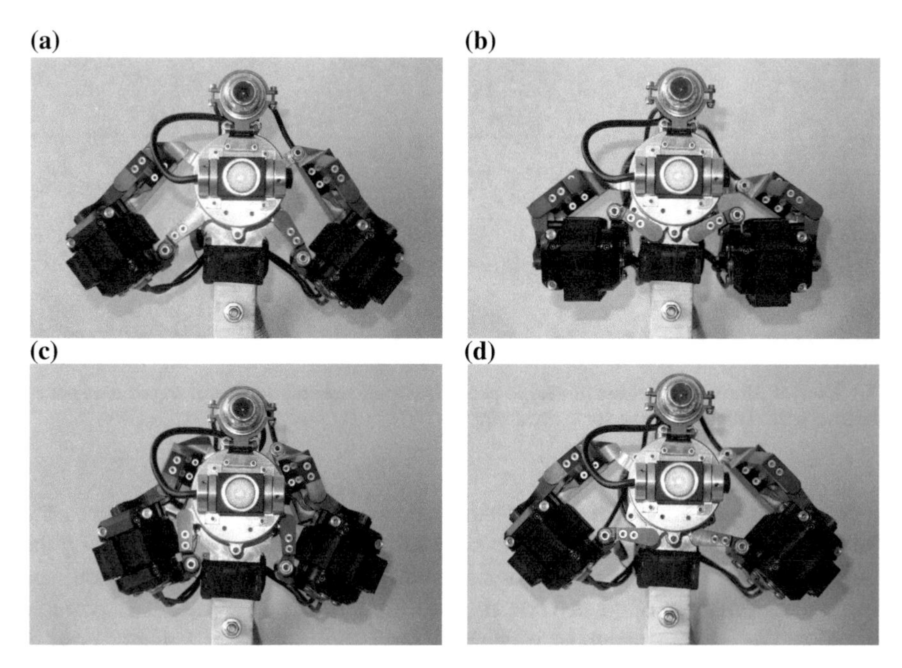

Fig. 7.6 Transformations of the prototype pedipulators—attained poses: **a** neutral; **b** for motion on horizontal surfaces or rectangular pipes and ducts; **c** for motion in Ø210 mm horizontal pipes; **d** for motion in Ø315 mm horizontal pipes

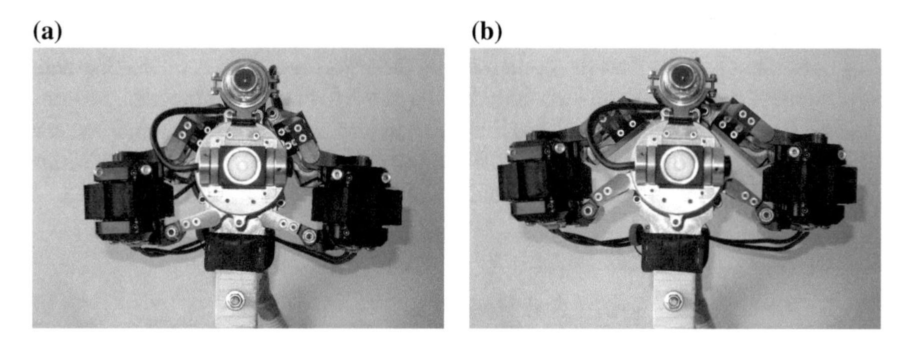

Fig. 7.7 Transformations of the prototype pedipulators for motion in vertical pipes: **a** Ø224 mm; **b** Ø270 mm

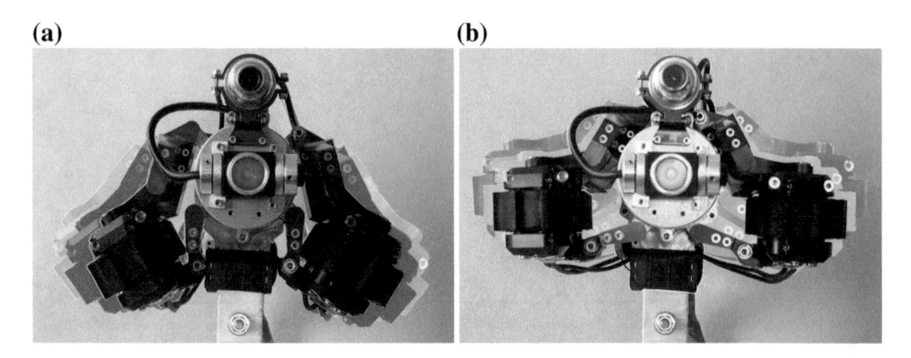

Fig. 7.8 Transformations of the prototype pedipulators—graphical overlap of selected trajectory steps: **a** for horizontal pipes; **b** for vertical pipes

Exemplary reconfiguration steps that can be realized in horizontal pipes are illustrated in Fig. 7.8a by graphical overlay of consecutively attained poses by the prototype, according to the trajectory. Similar graphical overlay of the prototype reconfiguration for motion in vertical pipes is shown in Fig. 7.8b. It is possible to realize the transformation in very small steps to fine-tune pedipulator pose to particular pipe size.

The prototype reconfiguration tests presented above prove proper implementation of the original pedipulator trajectory calculation algorithm and validate elaboration of the mechatronic system of the robot prototype, including all of its components. Additionally, it should be mentioned that operation of the prototype without intrinsically consistent joint trajectories, generated by the algorithm would violate kinematic constraints of the pedipulator structure and may lead to damage of mechanical components. It could include formation of backlash in joints, gear transmissions, because excessive forces would be exerted on all components. In the most severe conditions, pedipulator positioning servomotors and power electronics may be damaged.

7.4.2 Tests in Different Environments

After validation of robot poses, attained on a dedicated test stand that provides safe pedipulators reconfiguration, laboratory experiments were performed in environments analogous to desired work environment. The first test involved operation on horizontal surfaces. Figure 7.9a shows the robot driving on a floor, with camera and an additional light powered on. In Fig. 7.9b the robot prototype is shown in a horizontal pipe with diameter Ø250 mm.

The most demanding operating scenario of the robot was also checked. It involved inserting the robot to a Ø250 mm vertical pipe made of PVC plastic. The prototype pose was set to minimum track extension for vertical pipes (Fig. 7.7a) and then the robot was inserted in the pipe. Next, pedipulators poses were adjusted to extend

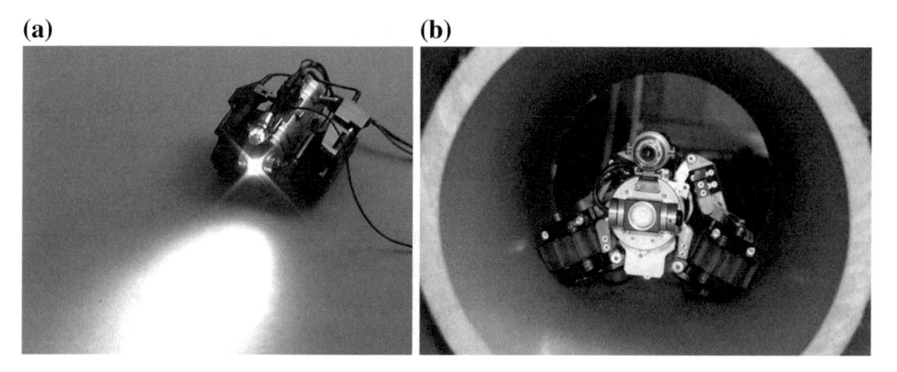

Fig. 7.9 Prototype of the robot in work environments: **a** a horizontal surface; **b** a Ø250 mm horizontal pipe

Fig. 7.10 Prototype of the robot in a Ø250 mm vertical pipe: **a** initial test of the tracks extension; **b** downward motion; **c** upward motion

the tracks and exert desired force on pipe walls. The extension was executed using previously calculated joint trajectories for servomotors. The prototype with extended track drives is shown in Fig. 7.10a. Two different robot orientations during motion in vertical pipe were tested. Orientation with the camera pointing downwards is depicted in Fig. 7.10b, whereas upward facing camera mode is shown in Fig. 7.10c. The test was performed in a dry pipe, where no drive stability problems were detected and the robot operated in a proper way. Operation of the prototype was tested for the entire velocity range of the track drives. If a wet or contaminated pipe would be used, potential traction problems may occur that could introduce limitations of prototype work in this demanding inspection environment. In this case, additional sensors are required to estimate track extension force and regulate pedipulators pose, according to changing surface conditions or geometric restrictions.

7.4.3 Experiments on a Dedicated Test Rig

A dedicated test rig was designed for better examination of robot motion and inspection capabilities. The test stand features an adjustable chassis that can be used to mount pipes with external diameter ranging from Ø200 to Ø315 mm.

Pipe mounting is realized by a rotary, adjustable support that can be set at an arbitrary inclination angle with 2.5° resolution. Thus, the test stand enables testing of the robot motion in pipes with changing inclination. It can also be used for running robot adaptation experiments between different pipe diameters. The robot prototype inside of the test rig is shown in Fig. 7.11. Video output from the CCTV inspection camera is shown on the monitor screen.

Experiment of the prototype motion in a Ø242 mm, dry vertical pipe, made of PMMA plastic is presented in Fig. 7.12, where we can distinguish three phases of the robot motion. It confirms that the robot can safely operate in pipes of any inclination.

Negotiation of elbows is an important capability of a robotic inspection system. The prototype during motion in a 90° pipe elbow is presented in Fig. 7.13. In this test scenario, manual track velocity setting with joystick control was used to adjust velocities to changing pipe curvature. It is possible to implement automatic elbow negotiation for the control system, but visual recognition of elbows would be required or preprogrammed track velocity curves may be used.

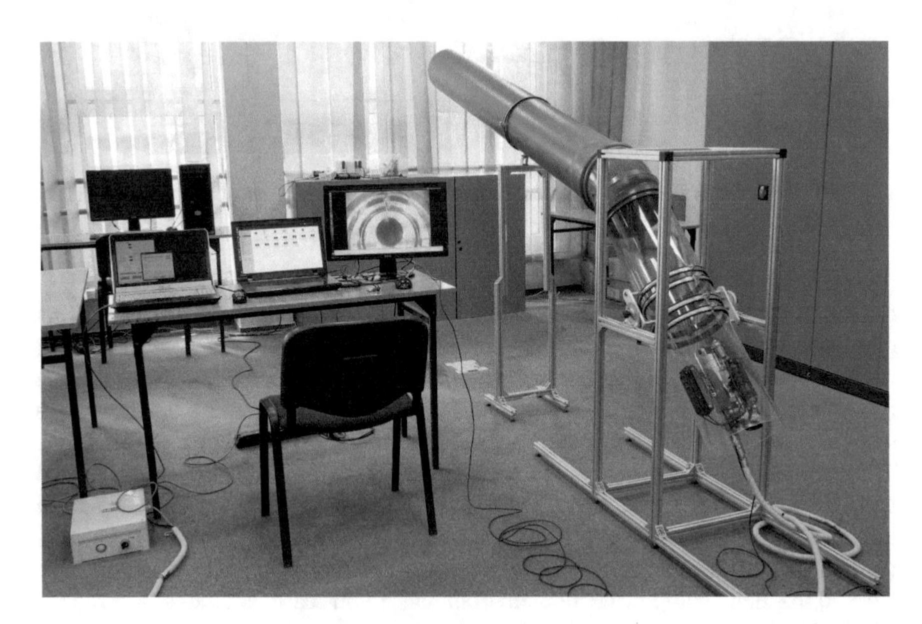

Fig. 7.11 Dedicated test rig for the pipe inspection robot

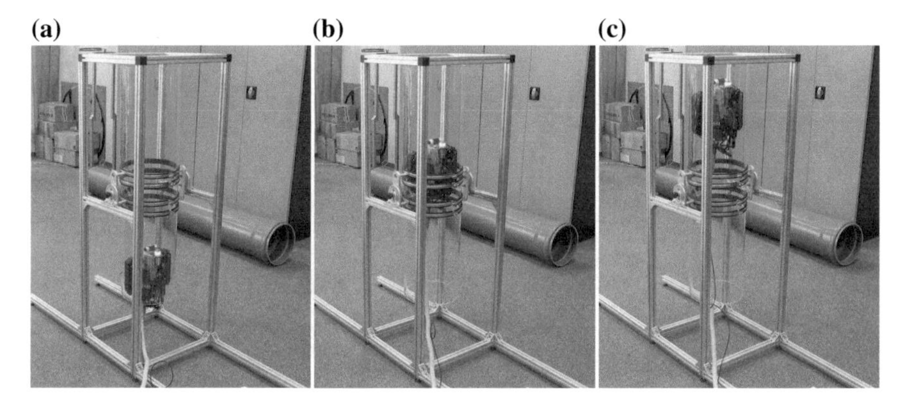

Fig. 7.12 Prototype of the robot during operation in a Ø242 mm vertical pipe—motion phases: **a** initial; **b** intermediate; **c** final

Fig. 7.13 Prototype of the robot negotiating a 90° elbow in a Ø300 mm pipe—motion phases: **a** initial; **b** intermediate; **c** final

7.4.4 Transformation of the Prototype in Pipes

On-line reconfiguration of the robot driving mechanism that can be performed in a pipe is an important ability and one of main advantages of the pedipulator design with active adaptation chassis, over simple, commercially available designs.

The reconfiguration is executed on the basis of pedipulator motion trajectories calculated by the transformation calculation algorithm. In Fig. 7.14, prototype reconfiguration steps are presented.

It can be noted that the transformation is possible also in the reverse order and for reducers that link other pipe sizes, with diameters available to traverse by the robot. Application limitations of this method are that the reducers should be oriented in a way such that bottom sides of linked pipes are tangent.

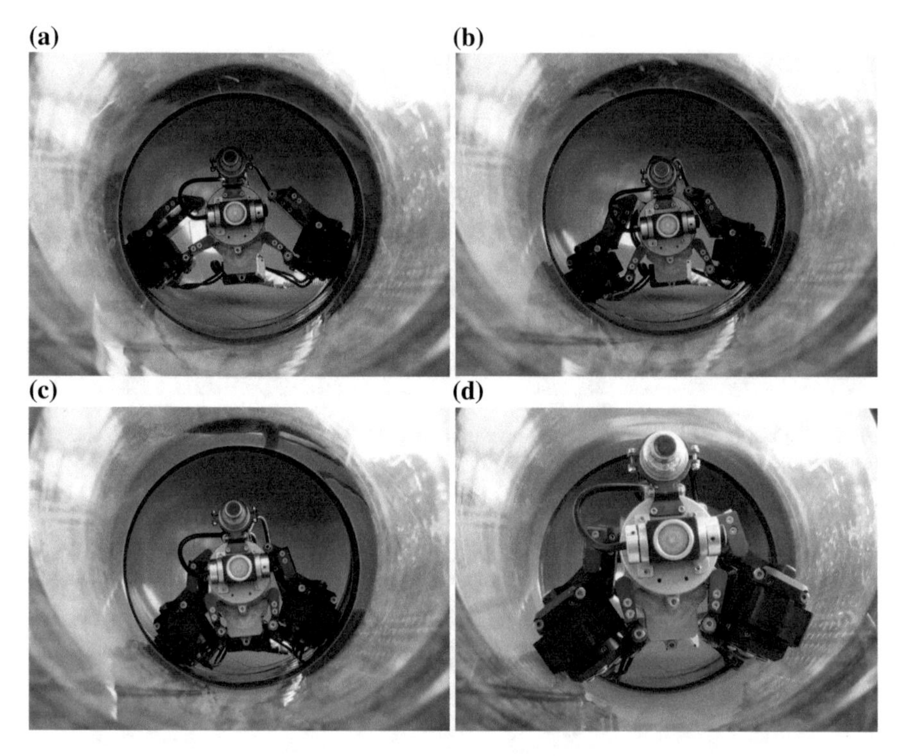

Fig. 7.14 Transformation of the robot prototype in a pipe with a reducer from Ø300 to Ø242 mm: **a** initial pose—Ø300 mm pipe; **b**, **c** intermediate poses; **d** final pose—Ø300 mm pipe

7.4.5 Laboratory Tests in an Environment Analogous to V-REP Simulation

For testing of the robot mobility and ability to reconfigure its track drives inside pipes, a test rig was prepared [1]. It consisted of straight pipes of nominal diameters DN315 mm (Ø300 mm) and DN250 mm (Ø242 mm), a pipe reducer connecting two pipes and two bends, one of diameter Ø300 mm and the second one Ø242 mm. They were assembled as presented in Fig. 7.15a. The objective of the test procedure was to verify if the robot is able to traverse the test rig. Initially, the robot was placed at the entry of the pipeline. Then, with operator manual control, attempt to pass the test rig was made.

For traversing of the rig, the following steps were prepared: drive into the pipe, traverse bend in larger pipe, get to pipe narrowing, reconfigure tracks to new diameter, get into smaller pipe (Fig. 7.15b), pass to next bend, traverse bend in smaller pipe, get out of the test rig. The performed experiment complied with results of the simulations and the robot completed the test successfully (Fig. 7.15c).

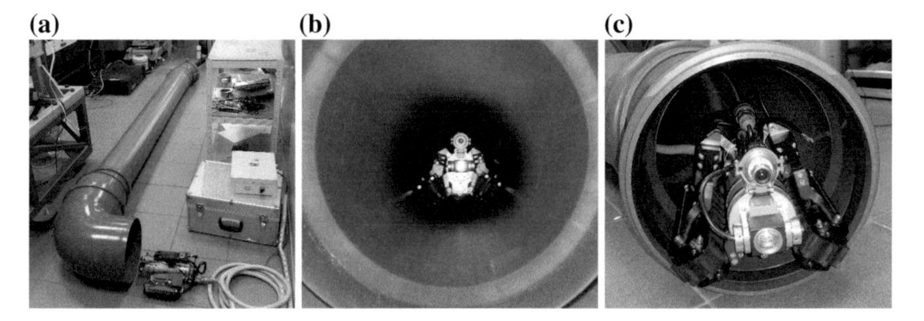

Fig. 7.15 Tests of the prototype: **a** the test rig; **b** the robot in a DN250 pipe; **c** the robot after completion of the test run

7.4.6 Measurements of Power Consumption

In order to estimate power consumption of the robot, the prototype was tested on flat surfaces (Fig. 7.16a) and in pipes with different diameters and orientations (Fig. 7.16b). Ten runs in each of three robot settings were performed to determine the minimum battery capacity, required for various work scenarios [2]. These tests can serve as an indispensable source of information for design of an autonomous version of the robot.

In the first setting, the robot structure was adjusted to operate on flat surface. The second type of environment was a horizontal pipe with diameter Ø242 mm, whilst the third experiment was performed in a vertical pipe of the same diameter. During all these tests, current drain was measured simultaneously for 24 and 6.4 V DC voltage power lines. The higher voltage source supplied the tracks and the lower voltage powered servomotors with the robot control electronics.

Data acquisition was performed using Intel Galileo development board, equipped with two Allegro ACS712 current sensors. Voltage measurements were realized with a universal multimeter Uni-T UT60A.

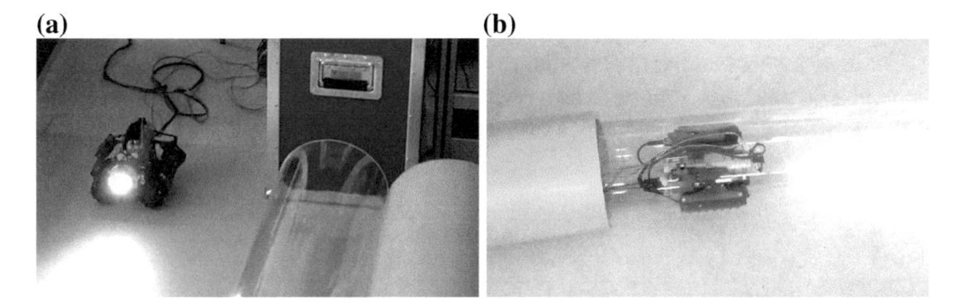

Fig. 7.16 Measurements of power consumption—test environments: **a** a flat surface; **b** pipes

In order to determine measurement uncertainties, error analysis was performed. The voltage measured by the universal multimeter can be determined for range $0 \div 40$ V, with the following limiting error (7.1) [3]:

$$\Delta U_{lim} = U_M \cdot 0.8\% + 1 \cdot R_D \tag{7.1}$$

where: U_M*—value of measured voltage,* R_D*—resolution of multimeter display.*

Limiting errors were calculated for measurements of voltages on power supply cables of track drives (7.2) and positioning servomotors (7.3):

$$\Delta U_{lim1} = 24.20\ \text{V} \cdot 0.008 + 1 \cdot 0.01 = \pm 0.204\ \text{V} \tag{7.2}$$

$$\Delta U_{lim2} = 6.45\ \text{V} \cdot 0.008 + 1 \cdot 0.01 = \pm 0.062\ \text{V} \tag{7.3}$$

The limiting error ΔI_{lim} that can appear during current measurements for the utilized sensors may be determined using Eq. (7.4) [3, 4]:

$$\Delta I_{lim} = I_{SR} \cdot E_{TOT} + \frac{\frac{U_B}{R_{AD}}}{S} \quad [A] \tag{7.4}$$

where: $R_{AD} = 1024$*—resolution of A/D transducer in the development board,* $U_B = 5\ V$*—input voltage range of the development board,* I_{SR}*—measurement range of particular current sensor,* S*—sensitivity of current sensor,* $E_{TOT} = 1.5\%$ *—total error of electric current sensors.*

For measurements of current flow in power supply cables, limiting error values were calculated using Eqs. (7.5) and (7.6) for track drives and positioning servomotors accordingly:

$$\Delta I_{lim1} = 5\text{A} \cdot 1.5\% + \frac{\frac{5V}{1024}}{0.185\ \text{V/A}} = \pm\, 0.10\ \text{A} \tag{7.5}$$

$$\Delta I_{lim2} = 30\text{A} \cdot 1.5\% + \frac{\frac{5V}{1024}}{0.066\ \text{V/A}} = \pm\, 0.52\ \text{A} \tag{7.6}$$

The errors were taken into consideration during calculation of required capacity of rechargeable batteries, proposed as a power supply for a future, autonomous version of the prototype. Instantaneous power was calculated with usage of mean values of measured voltages and currents, with application of the formula given in Eq. (7.7).

$$P_I = U_M \cdot I_S \tag{7.7}$$

where: P_I*—instantaneous power;* I_S*—measured current.*

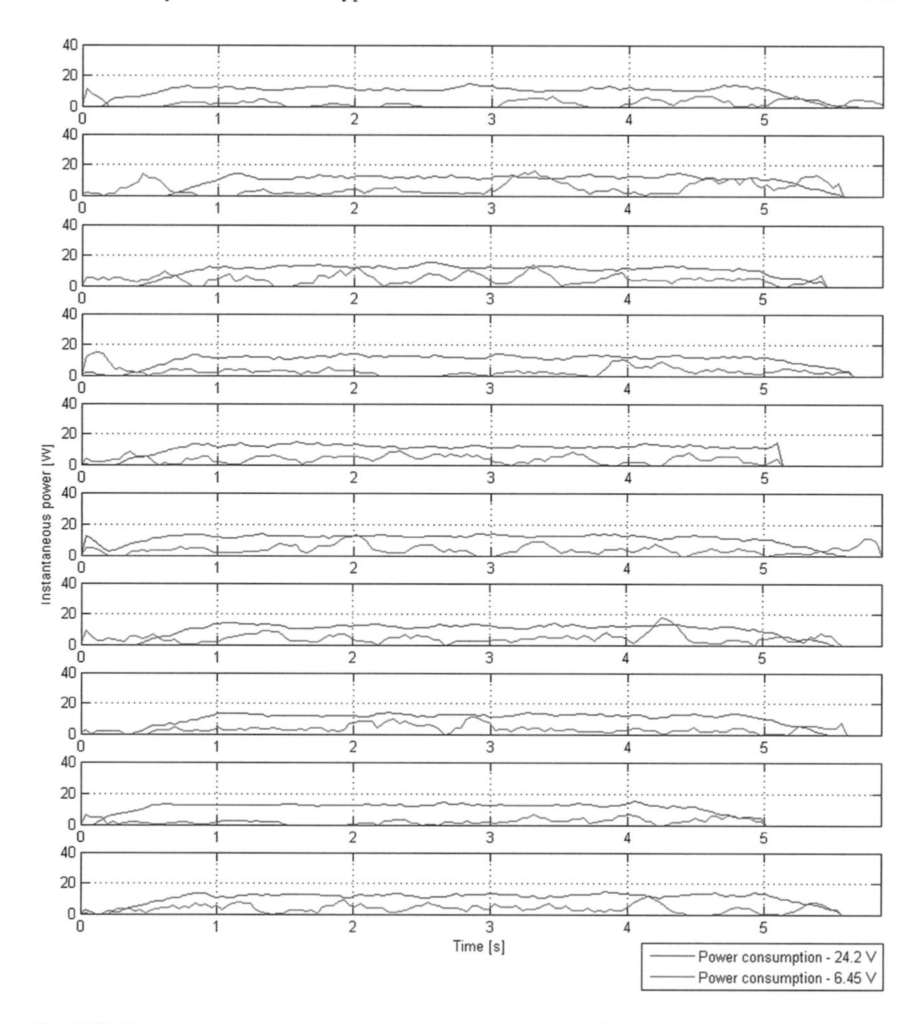

Fig. 7.17 Instantaneous power measured during 10 runs of the robot prototype in a horizontal pipe with diameter Ø242 mm

Instantaneous power consumption plots for track drives and positioning servomotors, measured during 10 test runs, are shown in Fig. 7.17 for horizontal pipe, Fig. 7.18 for horizontal surface and Fig. 7.19 for vertical pipe. Since current drain was measured on voltage 6.45 V for servomotors with control electronics and voltage 24.2 V for track drives, it is separately indicated with red and blue colors.

Results of the experiments indicate that the current drain during motion is below the operational parameters of the tracks, specified by the manufacturer. The difference of power consumption during motion in a horizontal pipe is within 10% margin with respect to operation on a flat surface, as indicated in Fig. 7.20a, b. During motion of the robot on a flat surface, large fluctuations of power consumption of

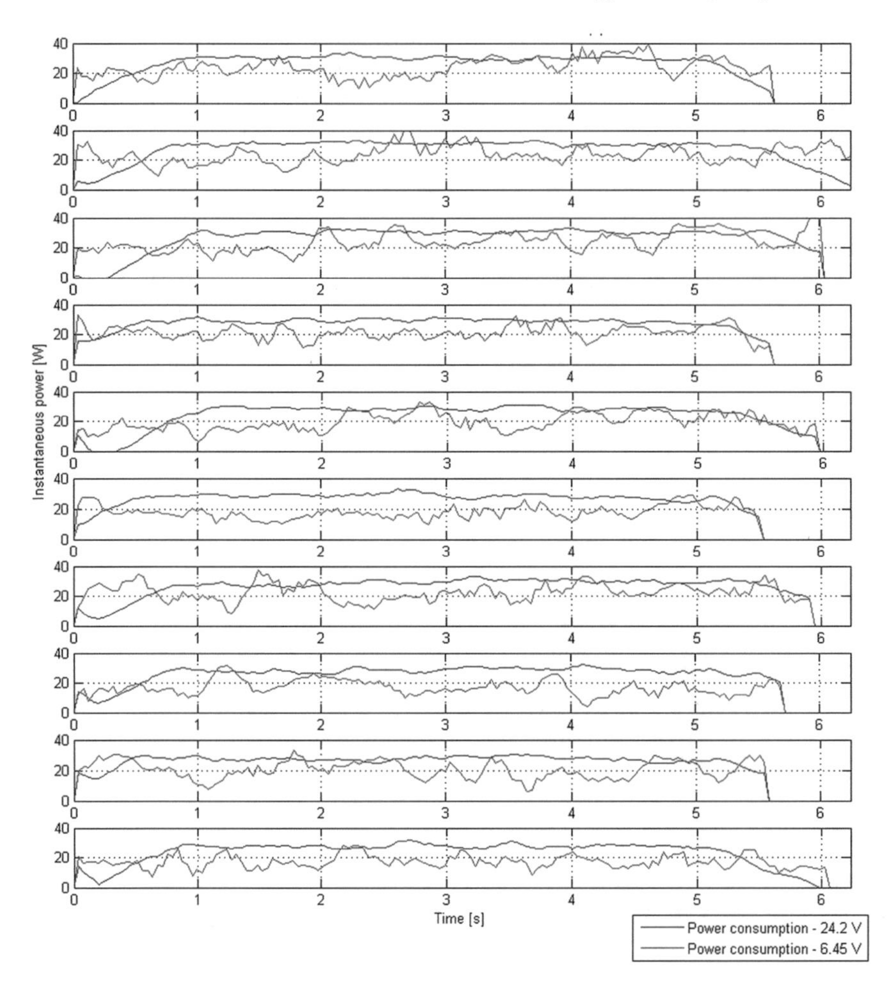

Fig. 7.18 Instantaneous power measured during 10 runs of the robot prototype on a horizontal surface

positioning servomotors (Fig. 7.20b) are observed. This is caused by non-uniform contact forces between the tracks and the ground. During operation in a vertical pipe, energy consumption is as expected the highest among all experiments.

In Fig. 7.21a, mean power consumption plots of the tracks for ten runs in each configuration are presented. It can be observed that during motion in horizontal and vertical pipes, the values are stable. During motion in a vertical pipe, power consumption fluctuates due to unsteady force exerted by tracks on pipe walls. In Fig. 7.21b, mean power consumption of positioning servomotors and control electronics is shown. Significant differences in values between consecutive robot runs for particular settings can be noticed.

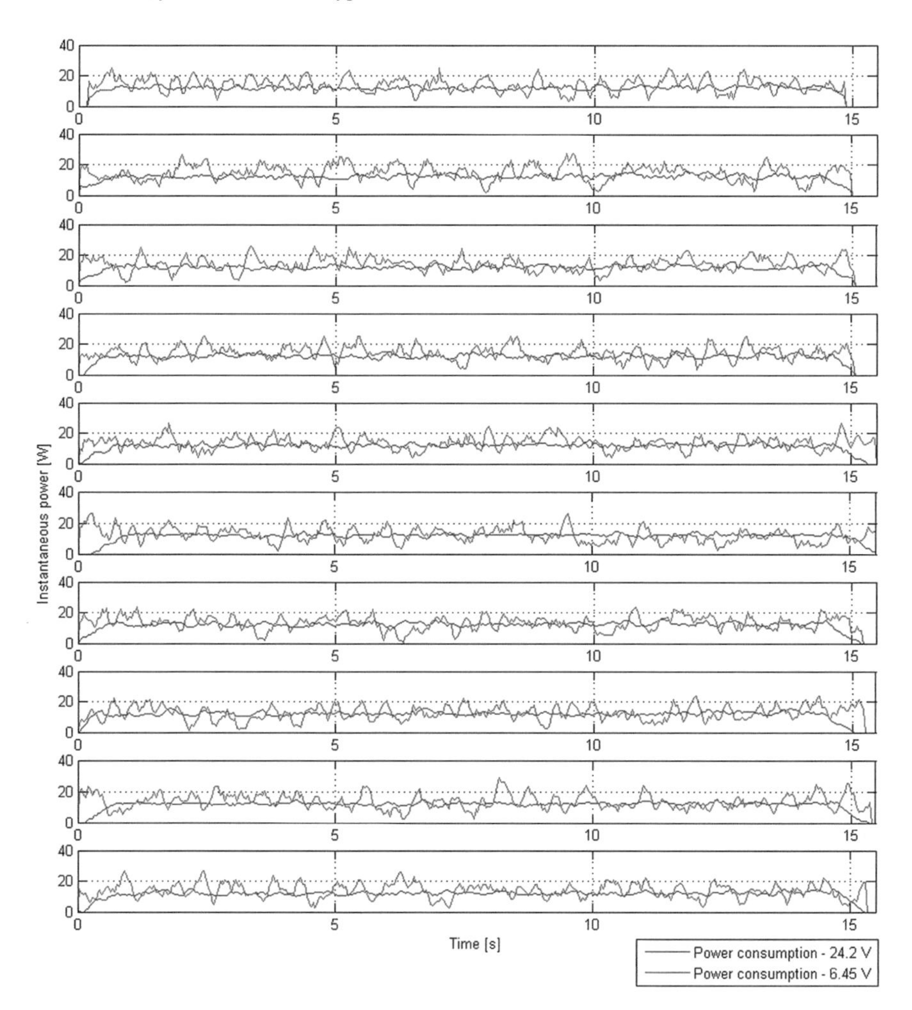

Fig. 7.19 Instantaneous power measured during 10 runs of the robot prototype in vertical pipe with diameter Ø242 mm

To provide better estimation of mean power consumption, normal distribution fits were calculated for all the experiments, taking into consideration previously presented measurements. Normal distribution curve fits for measured power consumption are presented in Fig. 7.22. The results for total mean power consumption of the prototype, calculated with this method have the following values, represented by maxima of normal distribution fit functions: 17 W for horizontal pipe, 26 W for horizontal surface and 52 W for vertical pipe.

Figure 7.23 shows an estimate of operating time of the robot on a battery power supply. It was assumed that the tracks and the positioning servomotors would be powered by separate lithium-ion polymer 2-cell and 6-cell battery packs, with nominal

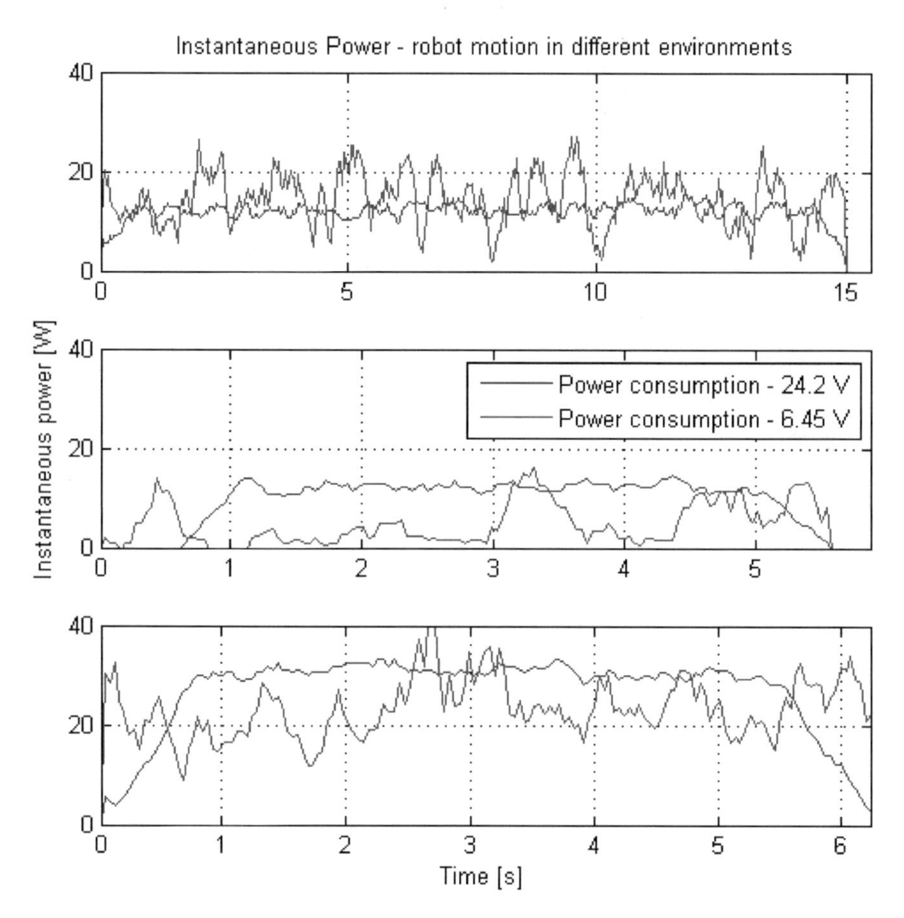

Fig. 7.20 Instantaneous power consumption in various operation environments—selected runs: **a** on a flat surface; **b** in a Ø242 mm horizontal pipe; **c** in a Ø242 mm vertical pipe

voltages of 7.4 and 22.1 V DC and nominal capacities of 3300 mAh. Total mass of the batteries in this setup would not exceed 0.9 kg. It would be possible to attach a proper battery compartment on the back of the robot body without significant alteration of its mechanical parameters and deterioration of motion capabilities.

According to data presented in Fig. 7.23, the robot should operate on the selected battery power supply for 3 h on flat surfaces, 6 h in horizontal pipes and 2.5 h in vertical pipes. This values would be sufficient for normal work scenarios during pipeline inspection.

The experiments showed that battery-powered version of the prototype can be built without excessive modifications of mechanical properties. However, to fully benefit from the battery operation mode, an algorithm for autonomous inspection would have to be developed and implemented in the prototype.

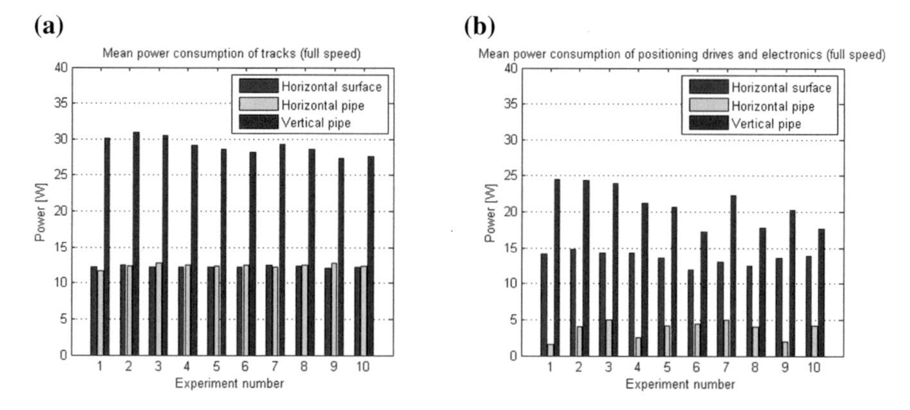

Fig. 7.21 Mean power consumption of the prototype subsystems: **a** tracks; **b** positioning servomotors and control electronics

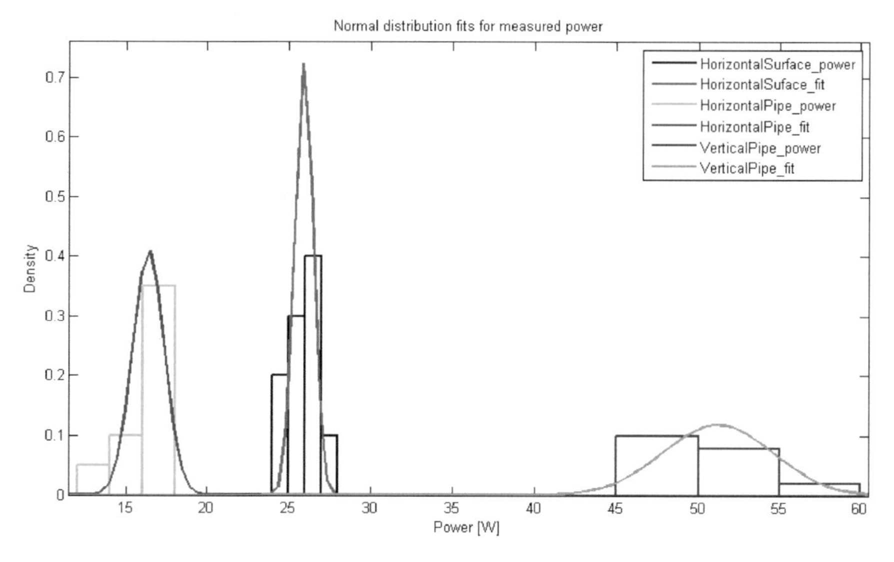

Fig. 7.22 Normal distribution curve fits for mean power consumption of the robot drives (from left to right for operation: in a Ø242 mm horizontal pipe, on a flat surface, in a Ø242 mm vertical pipe)

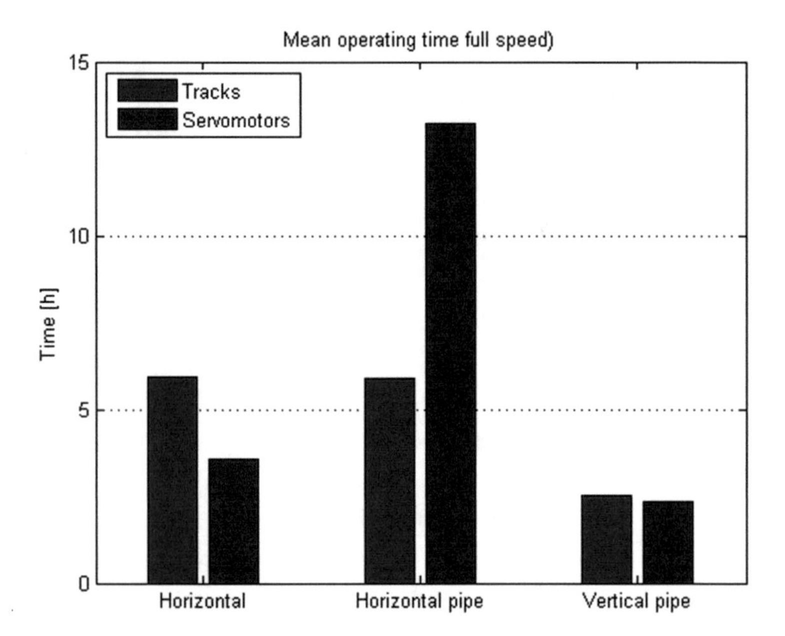

Fig. 7.23 Estimated mean operating time of the robot prototype on the selected battery power supply

7.5 Enhanced Version of the Prototype

On the basis of the experiments performed on the first version of the prototype, numerous improvements have been proposed. The modifications are mainly focused on the on-board controller, communication and power supply subsystems. Operation of the robot with usage of the Ethernet communication protocol was advantageous in terms of possible usage of high-definition IP cameras, standard network libraries and robust data transfer. However, application of an analog camera with sufficient resolution for investigated pipes with an integrated light source, eliminated the need of the Ethernet communication that requires multi-wire cables, prone to signal disturbances. The need of robot operation on larger distances, exceeding 100 m, led to selection of power supply with higher voltage. In the first prototype version, the 6 V DC power supply for servomotors was problematic in the aspect of large voltage drop that would appear on longer distances and a need of large cross-section areas of cable wires. Possible field application of the robot, without direct access to 230 V AC mains voltage, implied that usage of rechargeable batteries is a necessary option. Moreover, to estimate track extension force exerted on pipe walls, current sensors for all actuators are required. They would be indispensable for proper operation in vertical pipes with changing diameters and during autonomous motion through pipe reducers. These and other observations were followed by redesign of the power supply, communication and control systems of the mobile inspection robot.

7.5.1 Control System Upgrade

A scheme of the upgraded control system architecture, designed for the prototype is presented in Fig. 7.24. Operation of the robot by the user is possible in manual mode or in semi-autonomous mode, in which the robot adapts to predefined pipe sizes. Software implementation requires complete redesign, but its functionality is based on the one, presented in Sect. 6.6. The software, for the enhanced version of the prototype is developed in C# language and includes multiple improvements over its predecessor. The robot is controlled by the operator with usage of a wireless joystick connected to a PC or a tablet. Power supply of the robot can be either 230 V AC mains voltage or 12 ÷ 16 V DC voltage source e.g. from a rechargeable battery or a car 12 V DC socket. The power and communication subsystem consists of an AC-DC converter that transforms 230 V AC voltage to 12 V DC voltage. It is used when mains voltage is utilized for powering the robot. Next, a DC-DC converter is used to supply the robot. In the enhanced version of the power supply subsystem, 48 V DC voltage is used and a voltage transformer, located inside the robot body converts voltage for the electronics, servomotors, track drive modules and camera. Long operation distance and 48 V DC voltage implied usage of the RS-485 industrial

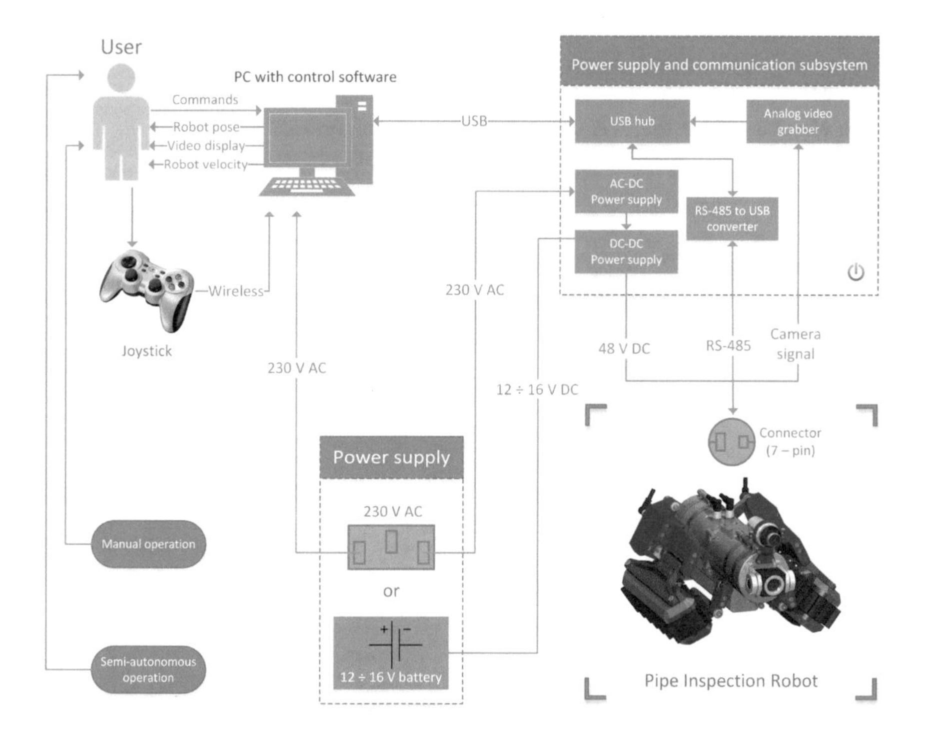

Fig. 7.24 Scheme of the upgraded control system architecture

serial communication system standard as an optimal solution. In the power supply and communication subsystem housing, a RS-485 to USB converter is placed and a USB analog video grabber is connected to convert analog camera signal to a digital one, prepared for computer processing. The robot is connected with usage of a Ø9.1 mm oil-resistant drum cable that consists of 4 pairs of 0.5 mm^2 cross-section wires. There are three options of the robot cable connections: short 10 m or 20 m segments for testing and short-distance inspections or a manually-operated drum with 105 m of cable for long distance operation. All connections with the robot are realized by robust 7-pin military-standard aluminum waterproof connectors.

The enhanced control system architecture provides better functionality in work conditions and minimizes necessary equipment to properly operate the robot.

7.5.2 Electronic Control Board

For building of an upgraded version of the prototype, a complete redesign of the PCB control board was conducted. The main requirements of a new on-board controller are presented in Fig. 7.25. The PCB was divided into two modules. The main controller is located in an analogous way, as in the previous prototype, in the bottom section of the robot body, whereas, the second PCB, a DC-DC voltage transformer is mounted in an additional housing. The voltage transformer feeds the track drive modules with 24 V and 2 A. The electronics are supplied with 5 V, whereas servomotors are powered by 7.4 V and 6 A maximum current, in contrast to 6 V in the previous version of the prototype. This higher voltage increases maximum torque and speed

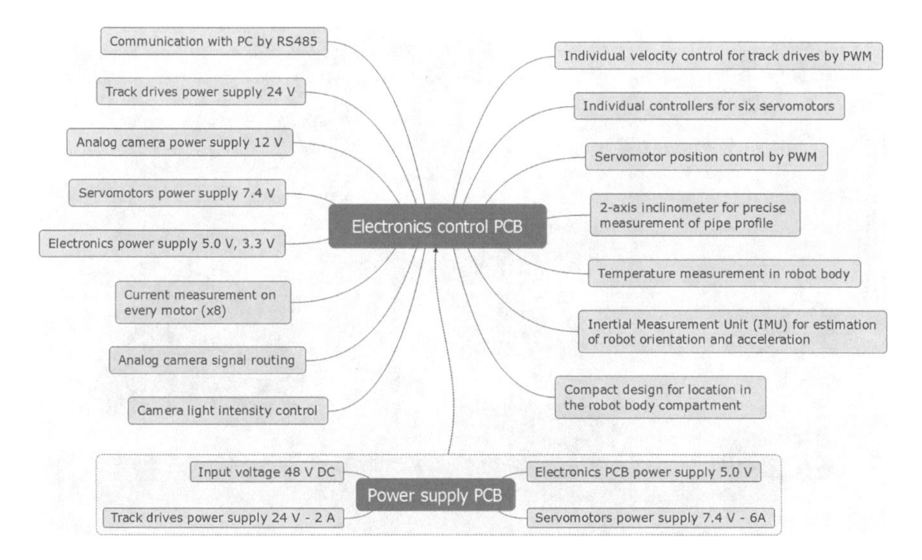

Fig. 7.25 Requirements for an improved electronics controller PCB and for a power supply PCB

of the servos that would operate in this case in high-voltage mode. Other supply voltages are generated by the main PCB board. The main changes of the on-board controller include installation of a 2-axis inclinometer for precise measurement of a pipe inclination profile. This inclinometer has 0.0035° output resolution and can provide data suitable for e.g. quality assessment of newly built pipelines. With usage of the IMU, mounted in the previous version of the controller prototype, such precise measurements were not possible due to noise, generated by high-range accelerometer. In the upgraded version of the control board, a 9-axis IMU is mounted for estimation of robot orientation and acceleration in workspace. It is particularly useful for motion in vertical pipes. Communication method with the PC computer was changed to RS-485 serial communication standard with the Modbus protocol, commonly used in industrial applications. One of the most important functionalities added in the new version of the controller feature current measurement on all robot motors. The current sensors will provide information about track extension state in vertical pipes and can lead to estimation of extension force.

Software for the electronic control board features all required functionalities, listed in Fig. 7.25 and was developed with usage of the C language in the AVR Studio environment. In addition to the mentioned changes, the controller is equipped with several safety improvements such as drive cut-off switches that can prevent overloading of servomotors at their stall currents. Another functionality is designed for restoring the last pedipulators pose attained before powering down the supply system. It is crucial during motion in pipelines that when the power supply is restored, the robot servomotors are not reset to the neutral pose that may be harmful for the mechanical structure and actuators.

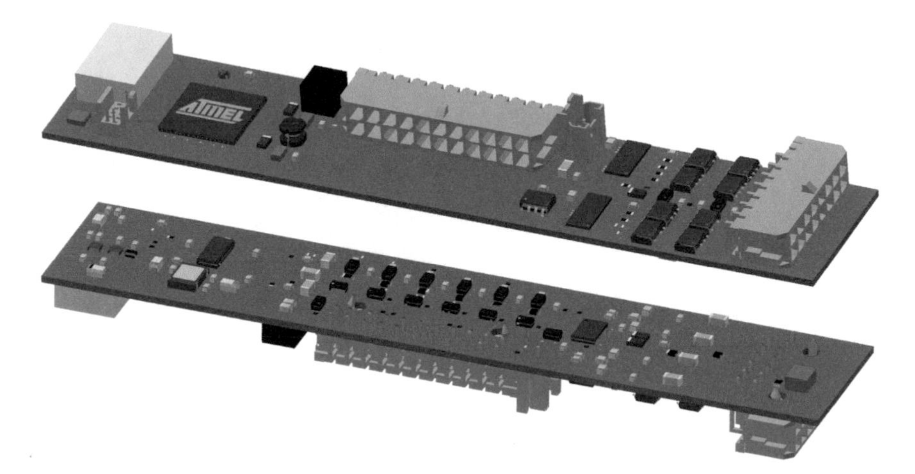

Fig. 7.26 The second version of the electronics control board—a 3D model

(a)

(b)

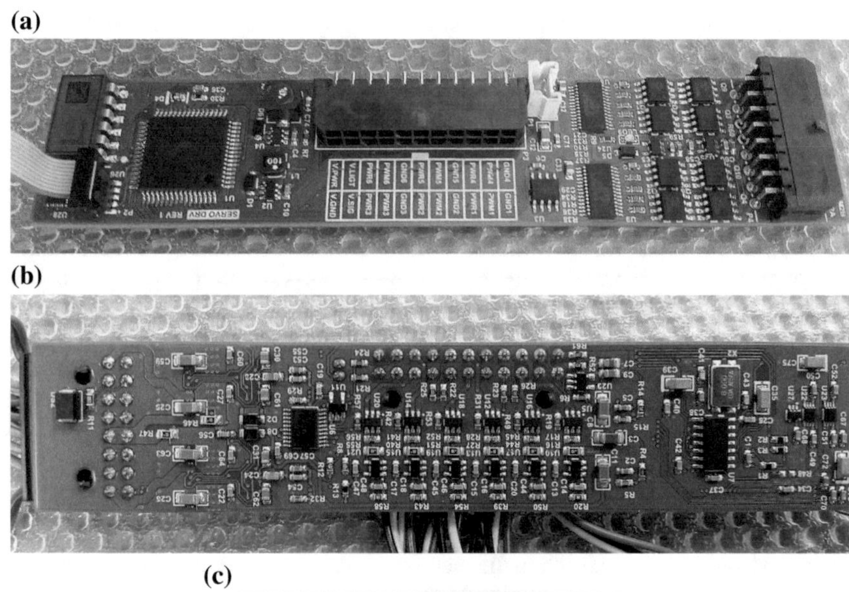

(c)

Fig. 7.27 Enhanced version of the control system—prototypes: **a** the main controller PCB; **b** the DC-DC converter and power supply PCB

A 3D model of the second version of the main on-board controller is depicted in Fig. 7.26. In this PCB board, AVR Atmel ATxmega256A3U microcontroller is used as the main processing unit that runs at 32 MHz frequency. Cable routing is realized by connectors that eliminate risk of unplugging after unpredicted robot impact.

The prototypes of the main controller board and the DC-DC voltage converter are shown in Fig. 7.27a, b respectively.

Due to high power and small dimensions of the voltage converter, laboratory tests were conducted at full load to determine heating of its subsystems. The main controller board was also checked to ensure proper operation of all components after robot assembly.

7.5.3 Upgraded Prototype of the Robotic Inspection System

For assembly of the upgraded prototype of the robotic inspection system, numerous operations have been done. The mechanical construction of the robot was changed to accommodate a DC-DC power supply unit and a new cable connector. The power supply, assembled in a new housing is shown in Fig. 7.28a. Cable connection with the power supply that provides tear and bend resistance is depicted in Fig. 7.28b. Moreover, a dedicated cable drum was manufactured to provide continuous data transmission independent on its rotational velocity. The cable drum was equipped with connectors mating with the robot's sockets and a slip ring that provides contact during unwinding of the cable as the robot progresses in a pipeline.

The power supply and communication housing was also modified and compacted. In the enhanced version, the AC and DC power supplies are hosted and a standardized robot cable socket is provided that allows plugging different cable segments or a cable drum. A USB hub that hosts RS-485 converter, analog video grabber and peripherals is included that can be connected to a PC computer or a tablet to run the control system software.

The full, assembled robotic inspection system for pipelines is shown in Fig. 7.29. The robot is plugged to the cable drum and the maximum operating distance is 105 m. The new control software is run on a tablet with Windows operating system and camera feed is displayed on the screen. Additional video output is provided, so that the video can be displayed on a separate monitor.

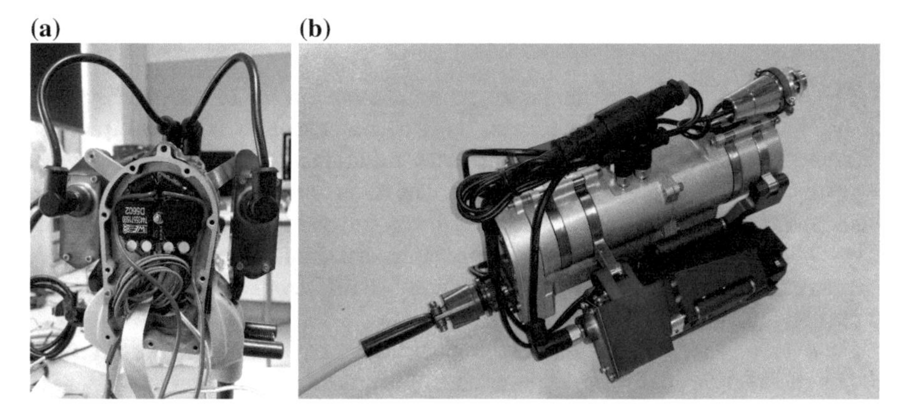

Fig. 7.28 Upgraded prototype of the robot: **a** the DC-DC converter in a dedicated compartment; **b** the assembled robot prototype with a new tether cable and an aluminum waterproof connector

Fig. 7.29 Enhanced version of the robotic pipe inspection system

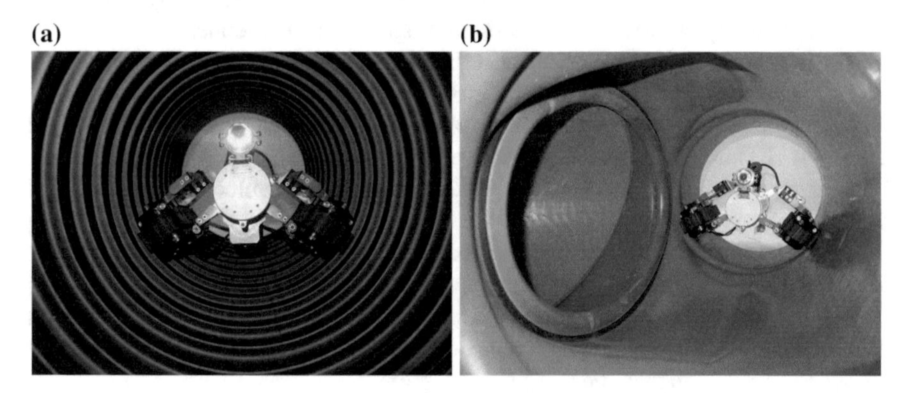

Fig. 7.30 Verification of the enhanced prototype: **a** a corrugated Ø315 mm drain pipe; **b** a DN315 T-junction—Ø300 mm internal diameter

The enhanced version of the prototype was verified in the laboratory. The tests included all previously presented work environments such as horizontal pipes of different diameters, vertical pipes, flat surfaces. Additionally, operation in other types of pipelines and junctions was investigated. The robot during motion in a corrugated drain pipe is shown in Fig. 7.30a, whereas the prototype in a T-junction is presented in Fig. 7.30b. The tests were successfully completed and the robot is able to negotiate corrugated pipe section and a T-junction of nominal diameter DN315 mm in addition to the previously mentioned work environments.

Final tests involved operation of the robot in water environment to check sealing quality and traction on slippery pipe surfaces. The robot during motion with extended tracks in a Ø250 mm horizontal pipe is shown in Fig. 7.31a. In this scenario, motion can be also realized in a vertical wet pipe. In Fig. 7.31b, the robot is shown in a

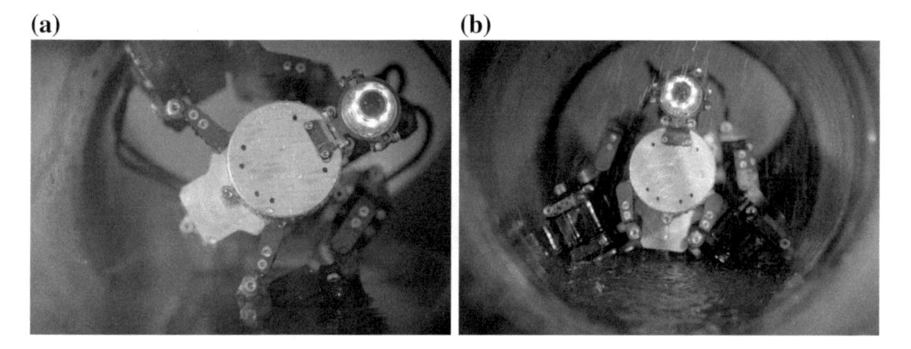

Fig. 7.31 Verification of the enhanced prototype in Ø250 mm pipe with water: **a** parallel track extension; **b** track adjustment for horizontal pipes

horizontal pipe of the same diameter but with normal track adjustment for horizontal pipes. Good traction was confirmed on wet surface and underwater.

In addition to the enhanced motion properties and improved robustness in harsh environment, the new software allows to realize much smoother transitions of the pedipulators mechanism due to higher communication frequency that improves overall robot performance, especially during adaptation to the work in vertical pipes and passing of reducers.

The complete pipeline inspection robotic system presented in this section is prepared for field tests and implementation in industrial applications for inspection of pipelines with varying geometry, vertical, inclined pipes and other environments such as hardly accessible places that require video inspection and visual quality assessment.

7.6 Summary

Prototype building process involved manufacturing and integration of mechanical components, electrical parts and electronic control systems. Implementation of low-level and high-level software was a part of the integration. During the entire development process, the prototype was compared with the CAD model. Electronic components were produced and individually tested with actuators and other components before final assembly of all robot subsystems.

Numerous laboratory experiments of the prototype were conducted. Initially, these tests included pedipulators reconfiguration assessment, validation of the mathematical models and the trajectory calculation algorithm. Next, robot motion tests in various environments were successfully carried out. Measurements of power consumption were also conducted to estimate required rechargeable battery capacity for the robot operation without external power supply. These tests were also used for development of power electronics for the second version of the prototype.

Finally, an enhanced version of the robot prototype was presented with focus on control system and electronic components. Improvements were introduced in the areas of communication, power supply, sensing capabilities, portability and operation distance of the robot. The prototype was tested in different work environments and the enhanced pipeline robotic system was prepared for industrial applications.

References

1. Ciszewski M, Giergiel M, Buratowski T, Modelowanie, symulacje i weryfikacja mobilnego robota do inspekcji rurocia?gów z adaptacyjnym systemem rekonfiguracji mechanizmu jezdnego. In: Projektowanie Mechatroniczne. 2016, pp. 17–26.
2. Ciszewski M, Waclawski M, Buratowski T, Giergiel M, Kurc K. Design, modelling and laboratory testing of a pipe inspection robot. Arch Mech Eng. 2015;62(3):395–408.
3. Uni-Trend Technology Ltd. UNI-T UT60A Operating Manual. 2014. https://www.manualslib.com/manual/538650/Uni-T-Ut60a.html. Accessed 8 May 2014.
4. Allegro Microsystems LLC. Current Sensor ACS712. 2014. http://www.allegromicro.com/~/media/Files/Datasheets/ACS712-Datasheet.ashx?la=en.

Chapter 8
Discussion, Conclusions and Future Work

8.1 Discussion and Conclusions

In this book, a robotic system for pipeline inspection has been presented. The research focused on design, modeling, analysis, control and prototyping of a mobile robot with an active adaptation mechanism that allows on-line adjustment to work environment. The major impact of this system in the field of pipe inspection techniques is the possibility to reduce number of devices, necessary to perform inspection tasks in various types of pipelines.

Literature and market research, covered in Chap. 2 presents current technological level in the field of mobile inspection robots, pipeline inspection techniques and dedicated devices for different inspection tasks. In the conclusion of this study, visual pipeline inspection is the most popular method and in-pipe inspection robots are devices demanded in application of different branches of industry. A versatile pipeline inspection robot with an adaptive motion unit would be an original solution that could optimize monitoring of industrial facilities.

In Chap. 3, an original, patented robot chassis design is presented. It features two independently driven pedipulators with closed kinematic chains, driven by six servomotors in total that allow to set poses of track drive modules used to propel the robot forward and backward. The pedipulators design consists of combination of passive and actuated rotary joints. It requires custom modeling and control approach for proper operation of the prototype. A complete, 3D model of the robot was prepared by the author with full technical documentation, necessary to manufacture a prototype.

In Chap. 4, mathematical models of robot kinematics and dynamics are derived. They are useful for inspection of inaccessible places outside of pipelines. These models are utilized to describe motion of the entire mobile robot on even surfaces, with focus on estimation of its position and orientation in workspace.

M. Ciszewski et al., *Modeling and Control of a Tracked Mobile Robot for Pipeline Inspection*,
Mechanisms and Machine Science 82, https://doi.org/10.1007/978-3-030-42715-3_8

The unique structure of pedipulator-based track drive positioning mechanism, induced the need to apply unconventional approach in mathematical modeling to achieve trajectories for different motion scenarios. In Chap. 4, an extensive description of the pedipulators mathematical modeling is given, with focus on the original transformation trajectory calculation algorithm. This algorithm uses data from the 3D robot model, forward kinematics, numerical inverse kinematics, analytical solution to inverse kinematics and numerous conditions to calculate valid trajectories for the closed kinematic chains. The algorithm is verified in MATLAB with usage of plots with trajectory characteristics. Modeling of the robot motion in pipes, based on forward kinematics and other equations, also helped in verification of the algorithm.

Recent advances in robotics simulation environments has opened possibilities for realistic verification and prototyping of control systems with usage of 3D models. Chapter 5 presents robot motion simulations and verifications of the algorithm with usage of co-simulations conducted in MATLAB, Simulink and V-REP environments. Integration of these software platforms allowed rapid, interactive design and testing of control algorithms before their deployment on a real prototype. Operation of the robot in horizontal and vertical pipes was verified and motion of the prototype on even and rough surfaces was successfully examined.

Design of the robot control system, covered in Chap. 6, required fusion of techniques utilized for mobile robots and arm-type robots. Due to generally restricted operation environments of the robot designed mostly for pipe inspection, the main issues encountered in mobile robotic systems such as navigation, localization and mapping of complicated environments could be partially neglected. However, precise measurement of pipe inclination and actual robot position had to be addressed. The most important aspect in the control system is related to pedipulators reconfiguration and adaptation to pipe size and shape. The control strategy relies on trajectory calculation algorithm described in Chap. 4. A custom on-board controller was developed for regulation of eight drives velocities and positions, reading of data from numerous sensors, capture of CCTV video camera signal and illumination adjustment. Control software that addresses low-level microcontroller architecture and high-level PC user interface was described.

In Chap. 7, development of the robot prototype is portrayed in context of mechanical construction, verification with the virtual 3D model, assembly of actuators, electronic components, integration with control system and software implementation. Prototype building process is followed by subsequent experiments that allowed verification of functional aspects of the system. They covered pedipulator mechanism reconfiguration and motion tests in different environments, including horizontal pipes of various sizes, vertical pipes, horizontal surfaces and comparison with simulation scenes, developed in V-REP simulation software. Energy management was addressed in laboratory experiments for development of a portable battery-powered system. Finally, enhanced version of the prototype design was presented with numerous improvements, mainly referring to control system architecture, on-board controller, extended usability and improved performance in terms of operation safety and future development.

The tracked inspection mobile robot, presented in this book is a mechatronic system that was designed and verified in the aspect of mechanical construction, mathematical modeling, simulations, control architecture, software implementation, prototype integration. The pipeline inspection robotic system, developed during this research is a ready solution that can be deployed for industrial applications in the field of visual inspection of inaccessible places.

Finally, it can be stated that the book of the doctoral book is fulfilled. The mathematical models were formulated for the tracked mobile robot, equipped with two track drives, control system for the active adaptation mechanism was implemented and a prototype was built and experimentally verified during motion in horizontal pipes of various shapes and sizes, in vertical pipes and on even surfaces.

8.2 Future Work

On the basis of design, development and implementation of the pipeline inspection robotic system, enhancements of different components could be applied.

Firstly, for operation in a wider range of pipe sizes, the system may be supplemented with replaceable arms and track mounts. It would make possible inspection of vertical pipes of larger diameters. Reduction of the robot size for motion in smaller pipes would require complete redesign, since it is mainly limited by dimensions of the track drive modules.

Secondly, the CCTV inspection camera system could be changed to a Pan-Tilt-Zoom unit that would provide more detailed view of the pipe inside surface. Yet, application of such camera would increase overall robot length and could reduce elbow passing capabilities. A supplementary, thermographic sensor could be used to enhance visual inspection capabilities. Other NDT sensors, such as ultrasonic transducers or MFL arrays could be mounted on the robot to provide more detailed information about pipeline condition and improve monitoring, dedicated for specific types of pipe networks.

Thirdly, the active adaptation system with regulation of extension forces, based on measurements from current drain sensors for the servomotors and the tracks might be further developed. This system would provide improved safety during motion in vertical pipelines with sediments, obstructions and other disturbances of pipe geometry by automatic adjustment of clamp force.

Another enhancement of the system can address improved navigation in pipe networks with branches. Application of a vision system, based on shadow analysis such as the one described in [1] would enable the robot to autonomously traverse elbows, T-connectors and reducers. A vision system could be utilized as well for automatic classification of pipe defects, similarly to the approach presented in [2].

An integrated control panel enclosed in a watertight, shockproof housing may be developed to improve usability in more demanding industrial applications of the robot.

Enhanced workflow for optimization of pipe inspection procedures could be achieved by development of a software package, integrated with monitoring database, linked with map drawing software in e.g. Geographic Information System (GIS). Additionally, automatic reporting module and other customer-oriented functionalities could be added.

Finally, in the aspect of mass production for large-scale industrial applications, the system should be optimized for usage of cost-effective and time-efficient manufacturing techniques, not applicable for development of a single prototype, presented in this doctoral book.

References

1. Lee J-S, Roh S, Kim DW, Moon H, Choi HR. In-pipe robot navigation based on the landmark recognition system using shadow images. In: 2009 IEEE international conference on robotics and automation; 2009. pp. 1857–1862.
2. Kohut P, Giergiel M, Cieélak P, Ciszewski M, Buratowski T. Underwater robotic system for reservoir maintenance. J Vibroengineering. 2016; 18(6):3757–3767.

Printed in the United States
by Baker & Taylor Publisher Services